HUDSON VALLEY

FOOD & FARMING

Why Didn't Anyone Ever Tell Me That?

TESSA EDICK

AMERICAN PALATE

Published by American Palate
A Division of The History Press
Charleston, SC 29403
www.historypress.net

Front cover, top, left to right: Chef Jean-Georges Vongerichten Beet Salad;
This morning's egg; Early peach from Fix Brothers Fruit Farm. *Courtesy
of Cayla Zahoran.*
Front cover, bottom: Hudson Valley Farm, Copake, New York. *Courtesy of
Stephen Mack.*
Back cover, top: Farmer Jeremy Peele of Herondale Farm and one of
the grass-fed British white heritage cows he raises on his farm in
Ancramdale, New York. *Courtesy of Cayla Zahoran.*
Back cover, bottom: Hudson Valley Tree Farm, Copake, New York. *Courtesy
of Stephen Mack.*

First published 2014

Manufactured in the United States

ISBN 978.1.62619.752.7

Library of Congress Control Number: 2014953170

Notice: The information in this book is true and complete to the best of
our knowledge. It is offered without guarantee on the part of the author
or The History Press. The author and The History Press disclaim all
liability in connection with the use of this book.

This book is dedicated to: My mother, Nancy Edick, whom I miss every single day and who made us all love, laugh, travel and eat. And my friends, who remind me every day how essential that is to life at the table and beyond.

To farmers globally—our "Starmers"—the real celebrities in food who wake long before we do in any weather, every day, with a commitment to best practices despite worry, lack of appreciation, little pay and security to make sure we eat fresh, nutritious food vital to life and local economies. We honor and thank you for working hard to feed us well.

Dynamic Tessa is a one-woman empire and is sure to be a name you'll know, if you don't already, very soon.
—Diane Clehane of Mediabistro

Tessa has so much energy and knowledge about the realities of our nation's food chain. She is a great ambassador.
—Dr Sam Simon, Farmer, Hudson Valley Fresh

As a homeowner in Upstate New York, if supporting FarmOn! helps get better food on more tables, then Tessa can count us in. A day of food, family and music with the added benefit of embracing our farm community...I'm proud to support FarmOn!
—John Varvatos, Designer

Congratulations! What you are doing is wonderful.
—Amanda Hesser, Author, Food Writer, Entrepreneur

Whether or not you support FarmOn!'s ideals, if you love food, you might be interested in the delicious lineup.
—Larry Olmsted, Forbes.com

Let's keep the conversation going to promote resilient agriculture in the Hudson Valley. Tessa is one of our most valuable resources, and I'm glad to be in her orbit.
—Bob D'Andrew, Local Economies Project, New World Foundation

FarmOn! Foundation (a non-profit) seeks to support family farms by (in part) getting young people fired up about agriculture and getting shoppers fired up about local produce.
—Jeff Gordinier, New York Times

CONTENTS

A Shaker-style gentleman's farm in Craryville, New York. *Courtesy Stephen Mack.*

for Maria —

Love the Lake —

and your Farmer

We do not inherit the land from our elders; we borrow it from our children.
—Amish saying

FarmDr ! my

Sweet friend

<3 tessa

ACKNOWLEDGEMENTS

At the table, there is no black, white, rich or poor—there is hunger and a need to eat good food. Good food is not a privilege.

To women everywhere who inspire and bring sustenance to the table to nurture us all with good food, conversation and love every day—thank you.

To Grandmother Dot, who made Sunday supper and farm-fresh food a reminder and connection to what is important in life. She inspired me to bring energy, laughter and glamour to life, no matter if you are in the city, on the lake or at the farm.

My heart and gratitude always to Joyce Varvatos for unyielding encouragement that it is possible and, no matter what they say, to follow your dreams—and your best friend's advice.

Thea, thank you for loving to eat food from the farm as much as me and sharing the virtues of FarmOn! with children everywhere who are also inspired by you. Love to you both and John for hugging us all and making eating only complete with music.

Love to you, Stephen Mack, for your genius looking beyond the surface, believing in me, sharing your family and generosity to help us all be happy, healthy and well.

Thank you to Eric Williams, for taking this journey in the country and unwavering support, reminding me food is love and loving me without judgment.

Big kiss to Marc Szafran for passion, patience, sensibility, truth and support.

• ACKNOWLEDGEMENTS

Thank you to John Bolton for honest guidance and safeguarding my direction.

Thank you to Paula Colarusso for reminding me every day we are more alike than different and without cooking, canning and crafting, we aren't living.

Thanks to Nanci and Chris Weaver for loving me like family and your kindness and thoughtful, insightful collective genius.

Love to my fabulous Tiara Club sisters Tiffanny and Charlotte for unending support.

And to my brother Kipp: you inspire me every day. I adore you, and you are perfect in every way. Thank you for all you do.

Without the support of each of you, the FarmOn! Foundation Board and these noble farmers—Ronny Osofsky, Ted Dobson, Jeremy Peele, Jim and Karen Davenport, Dr. Sam Simon and Linda and Bob Fix—none of this would be possible. Thank you.

Many thanks to Whitney Landis and Kathleen Willcox for your patience, support and hard work.

Grazie mille doesn't begin to express my gratitude to each of you ♥ (and Ruby Juice, too).

MY FOOD FORMULA AND CELEBRATING THE HUDSON VALLEY FAMILY FARM

My food legacy: Born and raised in Upstate New York, I fled farming life for a city mentality, a cosmopolitan lifestyle, success and glamour. I traveled around the world in search of education, a career and like-minded people. Funny thing about traveling in search of something is that everything you need is always within you, never where you are going. Ironically, I ended up back in an agricultural community—the kind of community that made me so happy in my youth.

My journey into food is unusual. I grew up with a mother who was vegetarian but told us when someone generously cooks for you, you eat what he or she made and say thank you. She actually was a pescatarian, now that I know that term—though I had no idea as a child. We just ate good food, sourced from the farm, and eating was social.

My great-grandparents lived on a farm with dairy cows, cornfields and a gorgeous garden. Each of us kids had our own personalized mason jar filled with sea salt that we took to the garden to consume fresh-picked veggies. I was the eldest of three children raised by a struggling single mom who somehow always managed to produce homemade meals made from farm-fresh, local ingredients for us every day. She created drive-through convenience with high-quality ingredients and homegrown love by getting to know all of the food producers within a stone's throw of our house. We ate with the seasons and frequented farmers' markets weekly for the freshest food. With grandmothers, aunts and cousins, we preserved and

canned those ripe, gorgeous fruits and vegetables at the end of each season at that farm so we could eat peaches, pickles, tomatoes and beans year-round. I had fifty cents to spend at each visit to the farmers' market, and my choices were always the same—fresh chocolate milk and fabric from the fiber farms I could use to make Barbie's clothes for her runway show.

My mother insisted that without breakfast, there was no point in going to school because you would be thinking about eating instead of figuring out how to change the world, so we were forced to the table in the morning for fresh bread and eggs or French toast from yesterday's loaf. Or Georgette, who minded me and whom I called "Gaga," would make me an egg (which she always got from the farm) over easy and tell me not to tell my mom it was cooked in bacon fat, which made it perfect with that crispy white edge. I loved to clean the bright orange yolk on the plate with warm buttered toast.

Mom packed our school lunch each day in the morning—always with a note of affection on our napkins that, at sixteen years old, felt anything but cool. Snacks were limited to fresh fruit and vegetables in Tupperware she would cut up and have ready to eat in the fridge. This was a far cry from the salty, sugary, crunchy treats our friends' moms served that made us all want more and an orange soda—the real forbidden fruit in our home.

By the time she was eight, my sister Tara was a professional sugar extortionist—she'd do anything for the stuff. One of her trademark bribes was picking our neighbors'

Right: This morning's egg. *Courtesy Cayla Zahoran.*

Opposite: The American silo and family farming in the Hudson Valley. *Courtesy Stephen Mack.*

flowers (straight out of their garden), knocking on their door and offering them their own flowers in exchange for an orange soda. Not my mother's proudest moment, but a family legend nonetheless.

Dinner was never at 5:00 p.m. like the families of five to fifteen kids that were common in our Irish Catholic neighborhood. I was constantly begging for cube steak at 5:00 p.m. like the Dermodys' meat and potato dinner, but since my parents were entrepreneurs and divorced, eating at "dinnertime" seemed to be an impossible feat. We ate closer to 8:00 pm and often out, mostly at restaurants owned by friends of my mom who fed us what they had cooked that day, something "special" that we would eat family style, which Mom loved. It was good eating and somehow—never mind her five-foot frame—my mom would eat her own food and off our plates too, a practice I still dislike today. Eat your own food from your own plate, and if you would like seconds, ask politely, please.

When we ate at home, it was always like a sitcom. When my dad and mom split up, my mom's friends stayed with us to help out—but they never left. I grew up with the "Ya-Yas" who were constantly in the kitchen and cooking and who made eating social and fun. We never had to eat anything we didn't like, but we had to try it once.

We never had to finish our plate, though we were encouraged to eat until we were satisfied, and we were confident that another meal was hours away. We were familiar with all types of vegetables, fruit, fish and freshly bottled milk. My mother loved to bake, and as a teenager, I followed her lead.

Since we grew up without money and both my parents were entrepreneurs, our eating schedule wasn't the norm, but there were a few constants. Milk came in glass half-gallon jugs from Byrne Dairy, and when we were little, it was delivered to the milk box off the pantry. We always had fresh tomato sauce that my mother would spend a whole day making in a special big heavy pot that started with scoring fresh tomatoes, removing the skin and seeds and melting into pure goodness for pasta, fra diavolo and other meals each week.

The other oddity at our table was the abundance of fresh lobster and clams. My mom had a client who imported and exported seafood and brought her a box of lobster tails weekly, which—when you don't eat meat and need to add protein to a meal—is a great go-to. We had more lobster

Opposite: At Trevor Valley Farm, freshly sliced tomatoes from the garden with tree-ripened peaches for lunch. *Courtesy Tessa Edick.*

growing up than you can fathom—in our eggs, in sandwiches for the beach, in pasta, in risotto, in salad, as appetizers and always as a perfect lobster roll, which is still, hands down, my favorite food today. It was ironic to always have the fancy shellfish as a staple in the freezer given our economic challenge, but I still appreciate every bite of every lobster I eat today and devour the sweet meat from every corner of the shell.

Speaking of shells, clambakes are another popular eating tradition in the Upstate New York summer months on holidays, birthdays or boating days. Friends and families would gather at lake houses of friends and family or tie up the boats on the lake and celebrate the few sunny warm days with laughter and food. We would shuck the clams and eat them raw with lemon and cocktail sauce or steam them with a quick dip in melted butter or even grill the cherry stone gems and eat them hot off their shells when they popped open. We ate dozens of them in a day with fresh butter sugar–style corn on the cob, and then we'd pair all of it with the one really unique food that no one knows about unless you are from the New York Finger Lakes region: the salt potato. It's a small "new" potato with a thin skin—not red and not fingerling—but when boiled with plenty of salt, it forms a crust and melts the interior into a mashed texture that never allows you to eat just one. Hinerwadle's was the farm that produced them, and I have yet to see them beyond New York State borders, despite the recent celebration of the one-hundred-year-old family farm.

Sunday dinner at Grandmother Dot's house was always formal. Your seat at the table started in the kitchen at the kids' table, and your manners earned your right to a seat at the "big table," ripe with witty conversation, good food and memorable eating habits with the family that sat ten or twelve any given Sunday.

At the "big table," less-than-perfect manners were intolerable, and the table was set with enough plates, forks, spoons and glasses to make Emily Post's head spin. If you weren't in training until your teenage years, you would be terrified arriving at the properly set table and trying to decide which utensil or glass to use when and how. But despite her ironclad rules, Grandmother Dot hardly ruled with an iron fist. She always pulled newbies aside and gave them one simple rule so they'd always know exactly which plate or glass to use for the right purpose. Make the OK symbol with both hands. The left hand forms a "b," and the right hand forms a "d." "B" stands for bread on the left, your bread plate, and "d" stands for the drink on your right, the drinking glass. Once we got those down,

A foraging, pasture-raised pig. *Courtesy Cayla Zahoran.*

we just watched and learned how and what to pick up next. It seemed somehow silly at the time, but I feel fortunate that my grandmother was convinced that "someday you will sit next to the president for dinner and your table manners will be useful!" Now I never worry when I'm at the table, whether I'm eating in the field with farmers or in a dining room with diplomats.

There were strict rules. If you burped at our table, you were immediately scolded and banished. You were to wash up, dress and come to dinner with interesting and informed conversation to share to make memorable meals worthy of the effort made to grow, prepare and cook the food. Because my grandmother worried that we would embarrass ourselves eating without decorum or politesse, once we had her rules down, her strictness actually freed us up to enjoy the meal and the company. It honored the bounty and the moment of a family sharing dinner together every Sunday afternoon and the goodness to come from that experience. My Grandmother Dot was right.

Holidays? They were madness growing up, even when our turkey was served with heaping sides of politesse. We were surrounded by two families that shopped, prepped, cooked and served everything you can dream up for a fattening celebratory meal. I had no idea how important it was at the time to source local and eat fresh, nutritious food because we always had it. Aunt Aurelia made everything we ate from local farms and fed us well. Thanksgiving was always crazed, as both sides of the families converged. We had everything from a microwaved turkey Aunt Laura insisted on making in the '80s (to show off *the* new invention she had first: the microwave oven) to a deep-fried turkey in the yard, which I think was an excuse for my uncles to build a bonfire and buy a keg. The one consistent thing was the turkey, which came from a local farm. Who knew that's why it was always so delicious, no matter the method of cooking or who made it. We were already thanking the farmer by buying direct!

We had a plethora of parties growing up. My mom believed that if you ate standing up, there were no calories. She always watched her weight living by these rules: never deny yourself any food you crave; eat fresh food in season, including plenty of fruit and vegetables; and three bites of anything is all you should ever need to be satisfied but stay fit and healthy. Our parties were always overflowing with people, whether it was a seated dinner or cocktail party with small bites. Sometimes the "FL" message was whispered around to us all, which translated

New York State's Hudson Valley is a food mecca and farming treasure. *Courtesy Stephen Mack.*

as "family last." FL was code for us to eat last because more folks showed up than we had prepared to serve! Never mind that Mom was a pescatarian—there was always a spiral-sliced ham roasted and served with mustards and chutneys and bread. Salty food meant people would drink, which Mom was convinced kept everyone laughing and made for a better party. She wasn't wrong.

Everyday eating varied but was always healthy, quick and easy, though it never sacrificed freshness, flavor or nutrition. If you eat food with nutrition, you eat less. You have fewer cravings and are generally thinner as you simply don't eat too much. Each week, Dad would come to dinner with us and the Ya-Yas, which meant a meal with at least five women, as well as delicious food of all kinds, chatter and conversation that was never dull and made me learn that the table is where emotion lives. There were tears and fights, admissions and corrections, disdain and love. These memories bring me back to the kitchen at night to cook. They bring me back to a family table to eat, and they forge a connection to the people who plant, grow and harvest the food we share.

Chatham Brewing on Main Street is committed to sourcing from the farm. *Courtesy Cayla Zahoran.*

We are all more alike than we are different. At the table, there is no rich or poor, no black or white, no labels on people being city or country. We all share the same need to eat and one common interest: nutrition to optimize the body and satisfy hunger.

The family that eats together stays together, which fosters values that last a lifetime, stability you can always revisit when life runs its course or runs you over and, most importantly, honors people who work hard to feed us so well: our farmers. When you eat local, you make a difference by optimizing not only your health but also the health of your community and create commerce by developing your region economically for sustainability, longevity and wellness. You live to eat—is there any other way?

Having traveled the world, I've experienced farming in many forms, but the commonality is the connection to your food. Somewhere along our food ways, we lost our way with food. We swap convenience for nutrition. We demand cheap food, and instead of giving value to our family, community and farmer, we let our ever-packed "to-do" list sideline family dinners for "activities." But being cheap with our food means we're cheap with our health—something we rarely value until it is too late.

Living in Europe convinced me that the age-old tradition of eating what's in season and what comes from the local terroir is sustainable and the only way to live. Eat strawberries or corn until there is no more left to pick and eat. Then switch to what's available next. There is no other way

ABC Kitchen beet salad. *Courtesy Cayla Zahoran.*

to get all of the nutrition your body craves and needs to optimize health and, ultimately our common goal, to look and feel great. You eat in moderation and in the rhythm of the seasons based on climate and yield, culture and availability. It's what all-natural eating should represent but has come to be meaningless on our over-processed, heavily distributed packaged food system.

Stand up for your farming community and local food choices and sources. Follow the seventy/thirty rule of eating local, organic and mostly plant-based food—not from an industrial plant, as Michael Pollan reminds us. If you opt out of processed food, you change the way you look and feel, and you don't obsess over the confusion about what is fattening or healthy or what the label says—or doesn't say. Naturally you will crave the food you need for good health and enjoy it without filling your body with empty calories that lead to a downward spiral in wellness throughout your life.

People always ask how I have so much energy and what do I do? I'll tell you what I don't do: I don't drink coffee like a religion, I don't eat processed food of any kind when I have a choice, I limit my sugar to organic raw and natural forms 90 percent of the time, I don't eat food without nutrition, I never eat anything from a microwave and if it isn't locally sourced and nutrient dense, I make a note to self: skip it.

Sleep, eat and exercise in the rhythm of nature, and your eating habits will form naturally. Make local farms your source for food choices instead of the tempting supermarket that isn't so super—where Madison Avenue is telling you what is healthy based on claims, buzzwords and sales goals from the big business of food. They do not care about the importance of eating good food, sustainability for the betterment of the people who eat it or the idea that a healthy lifestyle is the best preventative medicine.

We must weave our agrarian food ways and the locavore movement back to the table.

Take back control of what you eat—not with more TV recipes or supermarket trips but by making a connection to the farm. Any farm. As many as you can. Immediately. Forget who you are wearing and start only talking about who's feeding you because you care about how you look and how you feel. Too often, health isn't valued until it's lost, which is happening all too frequently with our fast, convenient now-now-now Big Food lifestyles. Who's your farmer?

But when you get down to the root of anyone's desires and cravings, health is the only thing we all really value. It's the basis for fertility and nourishment. For soaring joy, everyday happiness, sleep and connection with our fellow man. It's the only way you can act and live right. When you

Right: British white heritage cattle and Jeremy Peele of Herondale Farm. *Courtesy Cayla Zahoran.*

Opposite: Fruit you pick from the tree is always available down the road when you know your farmer. *Courtesy Cayla Zahoran.*

honor your body by eating quality food—ideally local, organic and honest food—you share a responsibility and accountability with the person who makes it for you and feeds you well. That balance is essential for the environment, the animals and the community, and it invites commerce to foster economic development for people, their families and friends. It moves you away from mass food production, cheap food and processed ingredients that leave you overfed and starving to death for nutrition.

When your food lacks nutrition, your body never optimizes and is never satisfied, which means you think you're hungry often when you already are full. When you eat food with nutrition, you feel satisfied, and you eat less. Your body isn't craving more.

The farm is also where you learn everything worthy of your commitment. The farm is a foundation of all that really matters, and the farmer is your conduit and is dedicated to feeding you well.

Eating this way sounds complicated, expensive, inconvenient or difficult. People always ask me "how to" FarmOn! In other words, how can they incorporate the fresh-from-the-farm mode of eating that is second nature to me into their harried lives that are already so over-scheduled? Beyond the diet, there is a solution and food formula to eating in the Hudson Valley. I've come up with food rules that will make farm-fresh food a part of anyone's daily plate. By utilizing modern technology and good old-fashioned agrarian values, literally anyone can connect with farms that grow our food and the legacy of immigrants and farmers who created a priceless heritage of real food that we are still reaping hundreds of years later. It's about eating nutrient-dense food that's healthy because it comes to your table locally, and that proximity—that freshness—is nutrition, and nutrition tastes better. It's an amazing opportunity to recognize, and I embrace it and want to share it with you. We are fortunate in the Hudson Valley, known as the breadbasket of the East; because of the richness of soil and the seasonal climate, production yields an agricultural surplus, which is often considered vital for the country as a whole. We also have sophisticated farmers and valuable academics and institutions. I'm grateful to have been connected to so many local farmers and food organizations all of my life, and now you can too—everyone can connect with the farm! It's a simple rule: meet your farmer. Everyone everywhere can follow this local food formula with food rules easy to implement beyond organic labels and fabricated health claims. Get connected to seed and soil. You are what you—and they (the animals)—

Berkshire pig at Sir William Farm. Redheads have more fun! *Courtesy Cayla Zahoran.*

eat. And like the chefs we celebrate, your farmer will show you the way from farm to your table.

So when you eat today, thank a farmer. He works hard to feed you well. He just needs more of us to buy from him directly. That small choice three times daily to eat quality fresh food is vital to our health, the health of our children and our communities and our welfare. Wouldn't you want to stand up for your farming community and food choices if you knew the upside? Ask: what is my criteria for buying food? How do my choices impact the family farm? Who do they benefit? How do I help preserve farming in America?

I want you to think about what you eat not based solely on taste and convenience but also why we must eat and base it on these three easy-to-remember thoughts that I will elaborate on later:

Responsibility: who made it?

Quality: where did it come from?

Nutrition: how was it made?

My mission is food education and farm preservation, meaning I will collectively bring a national awareness back to the family farm, highlighting the hard work of these Hudson Valley family farmers as a model. Hudson Valley farmers have persevered over the last few challenging decades of hardship. They are providing us with quality food that restores our common sense to make us demand food that is fresh, nutritious and supports our local community with commerce that fosters economic development. They are reestablishing a foundation of values and wellness that are pure goodness and make me inspired to bridge rural-to-urban marketplaces and inspire a re-education in food and eating that starts in public schools and with our children into what the "mother" of American food, Alice Waters, calls an "edible education." We can all re-create this model in our own communities.

It's super simple, really—the responsibility is in our hands, and we choose the sources. We, the people, have the power, not the corporations projecting our consumption with goals to feed the masses and poison us for profits with petrochemicals and lazy lifestyles promoting fast food with empty calories. We collectively must demand local food and relay this message to the farm by buying local and directing our food dollars to the family and small-scale farms. Our food choices create a better supply for everyone. In the business of commodity food, small choices like grabbing lettuce from a local farm versus grabbing a big plastic clamshell with lettuce sitting on the supermarket shelf for who knows

FarmOn! Ted Dobson of Equinox Farm, Tessa Edick, Ronny Osofsky of Ronnybrook Farm Dairy, Alice Waters and Jeremy Peele of Herondale Farm. *Courtesy Emmanuel Dziuk.*

how long and flown in from the other side of the country—or another country—make a big impact for a family farm, your community and your health.

All you need is food—and love. Food with nutrition is key. And nutrient density is vital for the body to optimize and function at the highest level of performance to keep your mind sharp, your body fit and your immunity strong. Nutrition is the buzzword that needs to trend in food. It should be your only concern when it comes to eating. Locally sourced foods that are fresh and come from the farm to your table mean you eat better. Eating food sourced closer to home, with more flavor, packed with vitamins and minerals you need for the body, means satisfaction, and the likelihood you overeat is less. When this practice of local-grown, farm-fresh, unmodified, pesticide-free, less-processed quality food sourced in the rhythm of the seasons starts at a young age with a connection to seed and soil, your habits for eating well last a lifetime. The connection to organic seed and uncontaminated soil will make your body a well-oiled

Friends of the Farmer at the FarmOn! Hootenanny! annual farm fresh benefit dinner, summer 2014. *Courtesy Emmanuel Dziuk.*

machine that will be in good shape for a long and healthy life. Is there anything lovelier? Yes, add fresh air and clean water!

This isn't a new concept. And it's an easier practice than you might think it is. We live in an agrarian society—we just (temporarily) lost our way from the farm. Back to the farm we must go. And when we do, and when we opt out of a system of poison for profits, we will mutually benefit ourselves and our communities. Our barns, our fields, our markets, our plates and our mouths will flow toward our overall wellness as individuals, and as communities, we will thrive and prosper. Isn't the switch to a collective consciousness in food shopping worth focusing on nutrition and having one conversation? It's an investment we all need to make and pay forward with prevention and better food. It's worth every penny and offers a windfall payback that not only tastes better but is also responsible and includes a commitment to honest food that goes way beyond the word "organic."

Chapter 1

HUDSON VALLEY'S LEGACY OF FARMING

The Hudson Valley is famous for landscapes, light and agriculture. It represents everything we love about America—hardworking people, fertile land, a river that roams from the countryside to the city yielding a breadbasket, innovation and fresh air that shapes the freedom we demand. The meadows, the hills, the fields, the orchards and the mountains bend along the Hudson River and inspire people to be responsible and cultivate honest food and explore the best the Hudson River Valley offers.

For four centuries, the Hudson Valley has served up some of the country's most coveted vegetables, fruit, grain, dairy, meat, poultry and spirits from heritage seed and fertile soil, and America has responded in kind, making it one of the most beautiful destinations in the world—a treasure along the Taconic—with delectable bites, breathtaking scenery and a pioneer spirit that is tough to replicate and impossible to fake. Like our European ancestors, the key to America's heart is our stomach.

Locally sourced, honestly made, farm-fresh food has commanded attention in the Hudson River Valley region again in the last decade. And not just that: it has reinvigorated farming with more farmers' markets, farm stands, community-supported agriculture (CSA) programs and farm-to-table restaurants that demand pastured meats, artisanal handcrafted recipes and food farmed with best practice in mind, be it organic, biodynamic or responsibly grown. The delicious changes have all been made with one goal in mind: to satisfy taste buds

and fuel our bodies with nutrition from the fertile land that connects us to seed and soil.

Food lovers covet the Culinary Institute of America, locally called CIA and located in Hyde Park near Rhinebeck, New York (home of Franklin Delano Roosevelt). The institute is wonderful, and the restaurant is a food Mecca, training in the culinary arts chefs who have notoriously become the most celebrated foodies in the world. Cooking with the best ingredients makes the best recipes and allows the CIA to practice sustainability and combine food ways from talented students and the local terroir.

The Hudson River Valley runs from the northern cities of Albany and Troy to the southern city of Yonkers along the eastern section of the state; it's nestled in the Appalachian highlands in communities bordering the 315-mile watercourse known as the Hudson River.

The river is named after Henry Hudson, who "discovered" it in 1609 (tell that to the River Indians, aka the Mohawk and Munsees who populated the valley in relative peace and democracy for generations before the arrival of the Europeans, who brought disease, fierce competition for

CIA students re-create Chef Jean-Georges Vongerichten's beet and yogurt salad with local organic ingredients in the Hudson Valley. *Courtesy Cayla Zahoran.*

Opposite: Hudson Valley landscape. *Courtesy Stephen Mack.*

Fix Brothers Fruit Farm in Hudson, New York, allows bees to cross-pollinate fruit with different types of trees mixed in the orchard. *Courtesy Cayla Zahoran.*

land and war). Native Americans fished in the river, hunted deer and game along its banks and cultivated the "three sisters": corn, beans and squash from seed.

The Hudson River slices through the middle of twelve counties, providing water and temperate maritime breezes that soften the sometimes harsh extremes of weather. During the Ice Age, New York was heavily glaciated, and when the ice melted, we were left with miles of rich, fertile (though rocky) soil perfect for an enormous variety of flora and fauna. Hudson Valley terroir can support row crops, vegetables, fruit, grazing animals and pollinating bees and butterflies. There is enough bounty and variety to feed all the hungry omnivores from New York City to Boston and back with delicious, seasonal food that's fresh, nutritious and sustainable—not shipped from halfway around the world with a carbon footprint as big as the entire Empire State.

Once known as the breadbasket of the country, the Hudson Valley is where we started growing our food as a nation. Before the first Dutch settlements were established around 1610 at Fort Nassau (just south of Albany), Mahican and Munsee tribes ruled the roost, planting fields to complement their steady diet of fish and game.

European settlers started arriving as early as the 1620s, and the French Huguenots followed forty years later, fleeing religious persecution and starting vineyards with their French winemaking skills, making the Hudson Valley one of the oldest winemaking regions in the country. (Turn to the resource guide for the names and addresses of my favorite Hudson Valley breweries, distilleries and winemakers, or Taste.NY.gov offers an extensive list, too.)

In the seventeenth century, settlers began building major colony operations. They claimed territory from the Delaware River to the Connecticut River and set up fortified outposts and colonial operations up and down the river. Fort Orange was the first permanent Dutch settlement in what was then known as New Netherland; it was located near present-day Albany.

Trading posts with farmers, craftsmen and laborers (including slaves from Africa) cropped up to feed, clothe and support the burgeoning communities. Small farms became the norm as America went through the growing pains endemic to the founding of an independent nation and major world power.

Centered as it was between New York and Boston, and on the banks of a mighty river, the Hudson Valley became ground zero for the British

Hudson River Landscape, by James McDougal Hart. *Courtesy of the Walters Art Museum.*

Opposite, top: A farm in Hudson, New York, circa 1941. *Courtesy of the Library of Congress.*

Opposite, bottom: Picking up corn after the reaper has finished at Mambert farm in the Hudson River Valley, near Cosackie, New York. *Courtesy of the Library of Congress.*

defense against the French invasion from Canada during the French and Indian War in the 1750s, not to mention a site for key conflicts during the American Revolution. (Taking control of the river, a source of transportation, food and communication, was a strategic and tactical maneuver both sides of every conflict tried to accomplish.)

The raging political conflicts and the reality of seventeenth- and eighteenth-century American storage and transportation capabilities made small farms work. Following the opening of the Erie Canal in 1825, which provided the first transportation system between New York City and the interior of the country, the population of the state surged, supporting small family farmers in a way that's unimaginable given how we operate today.

From the colonial era through the twentieth century, sprawling farms (with mansions to match) were also built by members of wealthy families with familiar noble names like Livingston, Van Cortlandt, Philips, Astor, Van Buren, Rockefeller and Roosevelt, among others.

East front porch of Montgomery Place. *Courtesy of the Library of Congress.*

Many of these homes and the remnants of their farms are still operating in some capacity, including the Philipsburg Manor in Sleepy Hollow, where visitors can see how a typical eighteenth-century farm was run and the labor (including slaves) responsible for growing the food that landed on so many people's plates. Check out my resource guide for

Montgomery Place. *Courtesy of the Library of Congress.*

information on historic Hudson Valley farms open to visitors, many of which also have seasonal farmers' markets and family activities.

In the eighteenth century, about 90 percent of the population in this country were farmers. Flash forward to the twenty-first century, and about 1 percent of the population self-identifies as a farmer.

In the three hundred years that ran between those bookends, the landscape that once supported the breadbasket that supported America has been developed, razed, built up and torn down. Innumerable technological, economic and sociological shifts have precipitated this radical shift in the manner in which Hudson Valley residents find employment, and I did not write this book to delineate or analyze those changes.

I wrote this book to celebrate and honor the farmers who are still managing to feed our hungry region without selling their principles, values or souls to the highest Big Agra bidder. These are folks who greet the crack of dawn with a smile, because they've already been up for hours

The Conovers at Sir William Farm in Craryville, New York, raise Black Angus cattle and other livestock with passion and a commitment to animal welfare. You can buy direct or at restaurants. *Courtesy Cayla Zahoran.*

milking cows, tending their pigs and touching the literal fruits of their labor in the fields.

Hudson Valley farmers are fighting an uphill battle: unpredictable weather and severe price competition inflicted by Big Agra and Big Box stores. In an environment of economic uncertainty with pressure from land developers, the region has been the tenth-most threatened agricultural region in America as far back as 1997. But at the turn of the twenty-first century, farmers started taking direct action to take back their fields and wallets, on their own terms.

Despite the wrenching changes of the last century, vestiges of the Hudson Valley's pastoral heritage have endured. Hudson was the first chartered city in the United States and is surrounded by trees of every kind. It is known for its apple orchards—of the heirloom variety—which are planted on the rolling hills that have yielded nearly half of all bushels produced every year in New York State for four hundred years.

The Culinary Institute of America, mentioned earlier and locally referred to as the CIA, offers several on-campus cafes and four-star restaurants that showcase seasonal, local, fresh-from-the-farm gourmet meals. Cooking with the best ingredients isn't a Hudson Valley invention. It's a European tradition. Michelin-starred chefs in Europe wouldn't dream of cooking with anything less than freshly caught, locally grown and carefully cultivated quality food. After all, it leads to the best taste. It also allows the CIA, which has been part of the Hudson Valley's local food heritage since its inception, to practice sustainability and superior combination of food ways from talented students.

A whole host of inspiring, grass-roots organizations are working tirelessly to restore the tradition of local food in the Hudson Valley. Some, inevitably, stand out from the crowd. The Hawthorne Valley Farm School and the Stone Barns Center for Food & Agriculture are paragons in the local food movement for the work they have done in the Hudson Valley and beyond for family farmers, local students and obsessed foodies everywhere.

In 2003, the Roundout Valley Growers Association was created to reinvigorate the health of farm businesses and launch an initiative for healthier school food. That same year, Tuthilltown Spirits founded the state's first legal whiskey distillery since Prohibition. While that may seem only tangentially related to farming, the fact is distilled spirits can increase

Tim Welly of Hillrock Estate Distillery explains flour malting and organic grain grown at the farm in Ancram, New York, where you can tour and taste their wonderful solera-aged bourbon and rye. *Courtesy Cayla Zahoran.*

the value of fruits or vegetables used in their creation by an astounding 800 percent—without sacrificing quality or adding chemicals, additives or preservatives. These people are not selling sameness!

Two years later, Hudson Valley Fresh was launched in a bid to secure premium prices for dairy producers. When it was launched, there were only 27 dairy farmers left in Dutchess County (down from 250 in 1970). Turn to Chapter 2 for more on their story.

Practicing localism and resiliency, the Local Economies Project of the New World Foundation supports vibrant, interconnected communities that are powered by the people who live there. They see food and farming as a cornerstone of every region, so their major initiatives revolve around food hubs (infrastructure and marketing), farm hubs (farmer training and services) and education (farm-to-school and community), and they are using a range of philanthropic tools to help these efforts take root and grow.

In 2005, Glynwood, an educational institution and sustainable farm, came out with a much-needed study that finally provided previously scarce data and statistics on Hudson Valley agriculture. Glynwood became the first go-to resource for farmers, legislators and business people who want to help create jobs and encourage sustainable farming, which not only feeds the region but also builds community character, attracts visitors (and their wallets) and preserves open land.

According to the most recent report (from 2010), if residents in the Hudson Valley and New York City spent just 10 percent of their food dollars on regional food, that would translate to a whopping $4.5 billion in food sales alone. That's the potential—about nine times the current yearly sales of all regional farm products. This just goes to underline my larger point: that putting money where your mouth is feeds and sustains so much more than just you.

The potential for supporting a responsibly grown food system is real, but the situation on the ground remains grim. According to Glynwood, in 2007, about 17 percent of the Hudson Valley (or about 1,325 acres) was devoted to farmland. Between 2002 and 2007, farmland acreage decreased by 10 percent, a trend that cannot continue.

"So what can I do?" is the common refrain I hear every day throughout my work building my own educational farm and working with chefs, legislators and food activists who are trying to promote a more just and sustainable food shed.

The problem seems so grand, so sprawling, so overwhelming. Don't get me wrong: it is. But the solution is surprisingly simple: meet your farmer.

Left to right: Hillrock Estate Distillery's Jeff Baker, Dave Purcell and Tim Welly. *Courtesy Hillrock Estate Distillery.*

Farmers are the real celebrities in food. They are our tried-and-true American heroes! They should be the heartthrobs on tractors we should be drooling over and begging for autographs from their food shows. These food stars I call "Starmers" hold the keys to our hearts, stomachs and food future with the health of our children and theirs in their hands and the hooves of their animals. The American farming family is the vista to our future. Support them every day by eating local. Get your FarmOn! Start today. It's the least you can do.

Chapter 2

IT'S COMPLICATED

You Are What You Eat

We already know that shifting our lifestyle and diet so that they work in harmony for health—instead of obesity and illness—starts with seed and soil. "Organic" has become a trusted buzzword in food that begs answers to questions we really don't yet know. The demand for organically grown food has skyrocketed since the USDA enforced national standards in 2002, totaling about $28 billion in 2012, an increase of 11 percent from the previous year, according to the Organic Trade Association. Overall, organic products account for more than 4 percent of total food sales in the United States. According to Nielsen, 55 percent of respondents were willing to pay more for a product if they know the producer is committed to responsible social and environmental policies (which is an increase of 10 percent since 2011, Specialty Food Trade reports).

We all know money talks. So as shoppers load up their carts with organic goods, farmers all over the world have responded in kind, producing more and more of their food with the best practices, ever mindful of the organic trend. But what if that organic lettuce is grown in China or California, and you live in Manhattan or Copake Lake? And what if that chicken you're about to dig into was fed organic grains but it lived out its life in a tiny dark crate and the organic grains happened to be genetically modified? Are the carbon dioxide emissions that lettuce leaf was responsible for producing or the lack of free-range pasture that chicken didn't roam worth the knowledge that it is organic, which means supposedly pesticide and chemical-free, if we assume bees, wind and butterflies were informed of the no-fly zone?

This is what they call rich-people problems, and the dilemma is often cast as an elitist perspective, but that couldn't be further from the truth. Our food choices present a very real, very modern problem that impacts environmental, social and economic change that affect every person in every walk of life every day on every level. This is about how we feed ourselves and our families, and it doesn't get more real or important than your food choices and sources of every single ingredient for every meal you eat.

When I was growing up, organic food was just food. My mom shopped for a head of lettuce at the farm stand down the street from us, and we never had to think twice about food miles. We knew that the heavier the head of lettuce, the better the value for a fixed price because we spoke with the person who made that lettuce head from seed.

We all have our food sins—mine is the "real thing" from a red can, even though it isn't even close to real food. I even buy the version from Mexico and tell myself the cane sugar makes it better. Occasionally, it's okay to indulge in things that aren't so good for us without building our entire lifestyle around the concept of drive-through convenience touted by food conglomerates.

"Pikkles" packed with local farm fresh food for a FarmOn! event. *Courtesy Cayla Zahoran.*

Obviously, it's complicated. Is organic healthier? Is local more nutrient dense? Is it possible to eat both local and organic without going broke? Is there any room in my diet for stuff that I know I shouldn't eat and drink but, quite frankly, can't live without—stuff I'm so addicted to that I keep eating without ever getting full?

Instead of making our lives more complicated, why don't we leave all of the complicated nutrition facts, percentages, formulas, claims, hypotheses and theories to the scientists cooking up their latest batch of petrochemical junk quasi-food at their gigantic, smoke-belching factories and keep it simple. Let's kick it old school, like we did when I was growing up.

I'm going to tell you what I do as a food entrepreneur, writer and philanthropist, activist and farm lover who has worked for and in the food industry with the most amazing foodies, retailers, chefs and farmers in America for over twenty years. I'll introduce you to a few of my favorites along the way for food shopping: I grew up with Wegmans, which I love; Whole Foods Market, my BFF; and Gourmet Garage, a dream on the island of Manhattan. Moreover, I'll let you know the most coveted and best-kept secret ingredient of the celebrated chef: the farmer.

This book isn't about diets, supplements or health plans; it's about visiting a farm for a walk in the fresh air with your friend and your kids and opening the proverbial barn door so that you can take a peek at the amazing buffet of options behind it stacked with responsible food, honest choices and nutrition so packed with quality you look and feel great. Pick and choose what works for you, and leave the rest for someone else to try. And pass it on. It works.

Eric Schlosser summarized our food state in his 2011 book, *On the Future of Food*:

> *Our agricultural practices are causing real harm. As Americans we are raised in different states and different circumstances but united by a belief that change must come, we want to reform the nation's current system of food production. It is overly centralized and industrialized, overly controlled by a handful of companies, overly reliant on monocultures, pesticides, chemical fertilizers, chemical additives, genetically modified organisms, factory farms and fossil fuels. Its low prices are an illusion. The real costs are much too high and they are being imposed on some of the poorest and most vulnerable people in the United States. Pesticides are poisons. The EPA has estimated that every year 10,000–20,000 farm workers suffer acute pesticide poisoning on the job and*

that's a conservative estimate. Farm workers, their children and the rural communities where they live are routinely exposed to these toxic chemicals. And what are the potential, long-term harms of the pesticides now being sprayed on our crops? Brain damage, lung damage, cancers of the breast, colon, lung, pancreas and kidney, birth defects, sterility and other ailments. Access to good healthy food shouldn't be reserved for a privileged few. It should be a basic right.

As more and more people demand not just organic food but also locally produced, farm-fresh food because of proximity, price and seasonality bursting with flavor, the market will continue to respond in kind, making healthy, whole foods available to everyone, not just people lucky enough to afford it, know the difference and demand decent food, no matter what the cost. We all have the right to good food. And just for the record, USDA-certified organic doesn't guarantee good taste or sustainable practices and is a certification program run by our government to market the word "organic" on everything you consume.

But we all have to eat. So what are our options?

I admit it: time is the truest luxury, and hitting the farmers' market just isn't always possible. The supermarket is not so super, and if you can, skip it. But if and when you must shop at the big business of food, stick to a few simple rules,

Right: Hudson Valley peaches tree-ripen, and farmers perfect the mouth feel and sweetness. *Courtesy Cayla Zahoran.*

Opposite: A picturesque Hudson Valley farm. *Courtesy Stephen Mack.*

and you'll still manage to leave with a cart of good food that tastes good too.

The simplest route back to our agrarian roots is to eliminate as much processed food as you can, without feeling deprived of whatever your canned or boxed vice of choice may be. Opt out of processed food and eat in the rhythm of the seasons, dictated by what our local Hudson Valley farmers are growing this week. The difference in taste is immediate, palpable and worth the switch, just for the sensual pleasure from consuming arugula grown within 25 miles and picked yesterday, instead of 2,500 miles away and picked two weeks prior. Never mind the environmental and philosophical halo that supporting my neighborhood farmer gives you. Food rooted in the terroir you walk over to get to your front door is conscientious, responsible eating that is sustainable and nutrient dense. Supermarkets these days almost always have a "local" section. Shop there and think about nutrition—don't judge a book by its cover!

A collective conscious consumption is vital to our own bodily health and the health of our children, but it's also essential to the health of our communities and our welfare as a society for commerce and economic development.

Wild goose chase with border collies. *Courtesy Emmanuel Dziuk.*

So before running out to the supermarket, try to remember these simple things and demand transparency. Instead of checking the expiration date, ask: when did it arrive?

Responsibility: who made it and with what methodology?

Quality: where did it come from and how was it raised, grown, cared for or fed?

Nutrition: how was it made and how old is it?

People get hung up on packaging instead of quality when it comes to their food. They pledge loyalty to brands instead of investigating how something is made and who made it. One of my favorite ways to pay farmers instead of advertising companies is to buy milk from Hudson Valley Fresh dairy cooperative (HVF). It's available in most New York markets, and if you haven't heard of it, it's probably because its advertising budget is literally zero. It's farmer owned and locally made in New York by family farms with the lowest somatic cell count and highest quality you can buy on the market.

Milk costs a lot to produce, especially if cows are grass grazing, happy, clean and lovingly cared for, like those from these ten dairies of distinction in Dutchess, Columbia and Ulster Counties. Imagine that instead of the valley of organic, you choose Hudson Valley Fresh for your milk needs, and your simple supermarket milk choice makes a huge difference in the lives of ten family farms. I know how the cows are cared for because I know the men who make the milk, and I know the cows are happy, clean and comfortable because their somatic cell count is 600 percent lower than the national average. (And you can learn this too—the farms in the HVF cooperative are all open for visitors. Go to hudsonvalleyfresh.com for details and addresses.)

Hudson Valley Fresh was founded in 2005. The dairy industry was desperate for change, and dairy farmers were going out of business as the price of producing milk was more than the price being paid by the commodity market.

Dr. Sam Simon, founder of HVF, was born on a family farm, and when he retired from his successful twenty-two-year career as an orthopedic surgeon, he committed full time as a dairy farmer and advocate of the family farm. As part of his mission, he decided to position HVF as a "Dairy Farmer Partnership" producing nationally awarded premium-quality milk annually. Today, HVF is also dedicated to preserving the agricultural heritage of the Hudson River Valley and promoting it as one of the premier food regions of the United States.

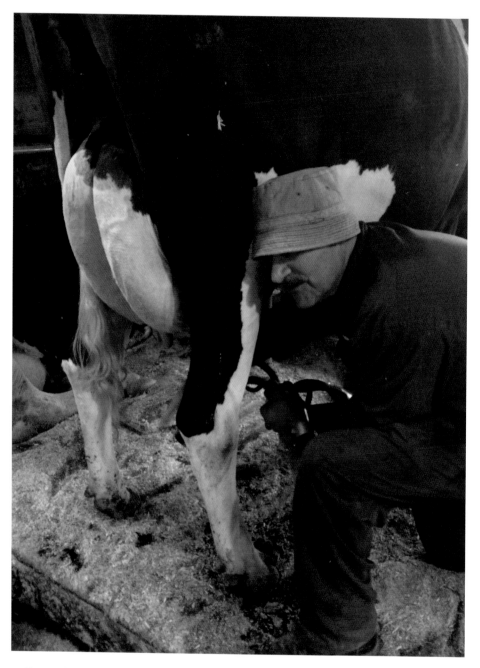

Farmer Fred Barringer at HillOver Healthy & Fresh milks twice daily the old-fashioned way with a strip cup and visual inspection for his home delivery market service from the farm. *Courtesy Tessa Edick.*

Based on a handshake, he purchased the 150-acre Plankenhorn Farm in Pleasant Valley in 1995 from one of his patients, Lester Plankenhorn, the only surviving member of the family, whose father had bought the farm in 1922 after crossing the Hudson River via the Poughkeepsie Railroad Bridge in a cattle car to keep the cows calm. The farm name remains Plankenhorn out of respect for the previous owners. "Great farmers, outstanding cow people," says Simon.

In 2003, Simon realized that receiving payment of around sixteen dollars per one-hundred-pound weight (sixteen cents per pound) for milk knowing it would cost nineteen dollars per one-hundred-pound weight (nineteen cents per pound) to produce that milk was impossible to do (even produced on a farm with no debt!) and survive as a business providing for a family and animals.

Simon explained, "There has to be a better way for quality production every day to earn a fair price. I was receiving awards for milk, as was Jim Davenport at Tollgate Farm, but the price paid for milk was still shy of what we deserved to be paid for premium milk as dairy farmers."

Knowing his loss meant other dairy farmers were losing their farms, Simon did something. He created a brand to sell premium quality milk based on worth, not commodity trading price, so the farmers cover their costs and earn a fair living wage. In 2006, the Hudson Valley Fresh dairy cooperative was formed and still today pays the farmers a set price so they can survive in business.

The Hudson Valley Fresh cooperative dairy mission is "to secure living wages for farmers and their families and ensure a fair price for our farmers' goods to keep those farmers in business," which means preventing the loss of their eight thousand acres of open land to development.

Simon told me, "Because milk pricing is commodity based, the price is determined on quantity of cheese sold—not how much milk is drunk or its quality—so it is a baseline for price. HVF, when it first launched, uniquely made milk, not value-added products like yogurt or heavy cream that command premium pricing. I knew quality milk deserved to be paid at least the cost to produce."

So Simon revolutionized the industry with his cooperative thinking and with Hudson Valley farm partners. They segregated milk by quality standards as a dairy cooperative—the only one of its kind in the nation that pays farmers premium prices for premium quality products and comes cow to carton in thirty-six hours.

Keeping farmers farming in family operations is vital to local economies in the Hudson Valley. *Courtesy of Stephen Mack.*

Cow to kid in thirty-six hours! Jersey cows lick a Hudson Valley Fresh milk carton. *Courtesy Joan Horton for* Country Wisdom News.

Most milk is at least a week old before you ever even see it on a supermarket shelf. Hudson Valley Fresh milk arrives to schools—and your food market—within thirty-six hours of milking. This means better taste, nutrition and value than any USDA organic milk on the market. Cow to kid in thirty-six hours! Beat that!

The HVF dairy cooperative continues to grow with now ten family farms located in Columbia, Dutchess and Ulster Counties producing milk that exceeds all standards for quality and nutrition: Bos-Haven Farm, Coon Brothers Farm, Walt's Dairy, Tollgate Farm, Stormfield Swiss, Shenandoah Farm, Domino Farm, Triple Creek Dairy, Dutch Hallow Farm and Simon's own Plankenhorn Farm.

They are located within twenty miles of one another and process their milk collectively at Boice Brothers Dairy in Kingston, a family-run business since 1914. The milk is never ultra-pasteurized and never uses artificial growth hormones (rBST/rBGH), which means it is fresher and offers more nutrition than even organic dairy, which is ultra pasteurized and shipped cross-country, taking weeks to land on your supermarket shelf.

Their award-winning Hudson Valley Fresh Dairy Holstein and Jersey cows are happy, comfortable and fed a varied diet of lots and lots of hay and healthy grains grown on the farm where they live, which makes a big difference in milk production. For Hudson Valley Fresh heifers, this is the farmhouse rule: more comfort + less stress = better-tasting milk! All milk is not created equal.

Dr. Sam Simon seized the opportunity to implement his vision for a better dairy future by ensuring that farmers are given a fair living wage price for milk they produce. With little more than a handful of struggling dairy farmers who just wanted to produce healthy, grass-fed dairy for local consumers, he has turned HVF into a regional favorite, winning many awards year after year for the premium quality delicious milk, not to mention a more viable livelihood for the farmers, humane life for the happy cows and money in the farmers' pockets for their hard work. This builds the local economies throughout the rest of the community, which benefits from having farmers who are providing jobs and drawing tourists to visit the farm, eat and pet their grazing animals and take in the views at many gorgeous, bucolic spreads.

When you buy Hudson Valley Fresh milk at the supermarket or ask for it in the coffee shop, you reward yourself and the hard work of ten family farms who produce dairy products: whole, 2 percent, fat-free and chocolate milk, as well as heavy cream, half and half, sour cream, yogurt and, just in time for the holidays, old-fashioned eggnog.

My personal favorite? Chocolate milk. It's the perfect recovery drink, the real energy drink and your kids' new favorite snack made with Pennsylvania Dutch cocoa and never any artificial ingredients or high-fructose corn syrup.

In fact, drinking local milk is vital to your family's health, the local community and economic development in the region. Buying Hudson Valley Fresh premium quality milk is not only a responsible food choice—it's also your community service!

In an industry where the cost of producing milk is much more than the pennies allocated from a commodity-traded product to pay the farmers making it, your simple supermarket milk choice makes a huge difference.

Speaking of supermarkets, why aren't they acting super by leading kids to the farm at their markets instead of marketing junk disguised as farm-fresh, nutritious food?

MOLASSES MILK (SINGLE SERVING):

Recipe by Paula Colarusso (@thefoodishgirl)

INGREDIENTS:
1 cup of fresh whole milk
1 tablespoon black strap molasses
1 tablespoon hot water

METHOD:
In a covered cocktail shaker, add a good portion of ice and then add milk.
In a separate ramekin, mix your molasses and hot water together—you do this so the molasses doesn't land like a blob onto your ice—doesn't have to be perfect!
Pour molasses mixture into shaker, COVER, shake vigorously, serve icy, cold and frothy—immediately.

BENEFITS:
Great source of nutrients and iron! Perfect for kids and pregnant mums.

ADD-IN OPTIONS:
Grate of nutmeg
Shake of cinnamon or pumpkin pie spice
Homemade hot fudge sauce (use the hot water technique)

If you want to do more with your food dollars and you're lucky enough to have a yard in the beautiful Hudson Valley, growing some of your own food is one of the simplest and most rewarding ways to know what you eat and get kids connected to healthy food. Planting a garden is easy to do, even for those of us who don't have time or a green thumb.

It all starts with a seed. Organic seed and organic soil are the very essence of who we are and how we should live to revitalize how we look and feel and our imprint on the environment, according to my friends at Fruition Seeds, who are helping to sow the seeds of a DIY gardening revolution.

"Organic seed is a huge growing industry just like the organic food market. Farmers and gardeners continue to adapt more sustainable and ecological practices and connect to how their growing practices impact not only food but air, environment and community," Matthew Goldfarb of Fruition Seeds in Upstate New York explained. I had painted my red-bottomed shoes black and gone back to the farm, but I still needed a confidence boost to fully resurrect my farming heritage and implement my cultivation skills for effective yield, timing and practice—talking to the seed farmer made sense.

What type of seed to plant when, how deep and in what soil was also as confusing as the science of animal feed, so I talked to more seed specialists and quickly learned that while gardening is simple, a seed is far more complex. The only way to ensure that your seeds will biodynamically air pollinate and organically produce food is to know your seed-saving farmer and where the seed came from.

Turtle Tree Seed Company and Fruition Seeds are both organic seed companies growing, breeding, saving and developing varieties for growers in the Northeast from seed in the same region. They are curating and customizing organic seed specifically for us so that we will have varieties that thrive today, tomorrow and for future generations, specifically for our climate. (No, avocados and citrus won't grow organically in the Hudson Valley, no matter the effort).

Goldfarb is the co-founder of Fruition Seeds, which he started with Petra Page-Mann in 2012. All of their seed is certified organic and open-pollinated. Fruition Seeds grows and saves seeds from sixty-plus varieties on its farms in Naples and Branchport and sources sixty varieties of seed from its network of other skilled organic growers throughout the Northeast, offering transparency about where and how seed is grown—an anomaly in seed breeding today!

The purpose of building this network of the best organic seed growers in our region is to increase the vitality of life itself. The alternative to corporate control of the seed supply is respect, knowledge and transparency and is based in regional seed breeding, selection and production.

Seed produced within its own bioregion is uniquely able to adapt to the varying conditions and demands of climate change, soil, pests and disease. Most seed around the country is sold by companies with national/

Opposite: Freshly harvested organic garlic. *Courtesy Cayla Zahoran.*

Left: Fresh organic garlic from seed. *Courtesy Cayla Zahoran.*

Below: Bucolic Hudson Valley farm. *Courtesy Stephen Mack.*

international markets that favor "widely adapted" seed production, so you may be buying good organic seed but not seed selected to excel in your specific climate and soils, so yield will decrease.

Developing strong regional seed is the foundation of healthy, vibrant and resilient food systems. Petra Page-Mann said:

> *Saving seed conserves the genetic diversity of garden vegetables—that has already eroded 75 percent in the last century—and ensures future generations will have diverse, resilient seed to grow in their gardens. Even if you do not save seed, buying open-pollinated varieties supports the diversity, abundance and significance of our gardens. Organic plant breeding can improve yields by adapting plants to out-compete weeds, be efficient at nutrient uptake and resilient to climate variability. Research from Washington State University showed that organic breeding can increase yield by as much as 31 percent. You can save true-to-type seed from open-pollinated but not from hybrid varieties. Unlike hybrids, open-pollinated varieties cannot be patented and have the ability to be suited specifically to your garden needs (like saving seeds of slow-bolting lettuce or non-cracking tomatoes).*

Page-Mann has great insight about the seed crisis in America:

> *Genetic patenting (including hybrids and GMO) has turned seed into "non-renewable resources" for farmers and gardeners who can no longer save their own seed. Each time we lose a variety of a seed, a food crop becomes extinct. As crop diversity diminishes the ability to adapt to climate change, pests and disease become more limited. Very few seed companies sell seed grown and saved in the Northeast. Those that do offer only a fraction of regionally grown seed; the rest is purchased from around the world. These seeds are produced for "wide adaptation" to grow "good enough" in some regions but may not excel in any of those places, especially the Northeast.*

The lack of breeding for organic and local farming needs is due to concentration of ownership within the seed industry. Did you know that only five companies control 60 percent of the global seed market? These companies focus on developing food crops dependent on petrochemicals, toxic pesticides, herbicides and fertilizer. How is that healthy?

Certified biodynamic is the name of the seed game, and Lia Babitch and Ian Robb are pioneers continuing to learn from and love their work teaching us all about better food sourcing at Turtle Tree Biodynamic Seed Initiative.

Co-managers of Turtle Tree Seed in Copake, their nonprofit seed company grows and sells 100 percent open-pollinated, non-GMO vegetable, herb and flower seeds, all grown using biodynamic and organic practices. They grow several dozen varieties on their farm at Camphill Village and redistribute seed from other biodynamic seed growers from around the country.

All of these seed curators are passionate. When Goldfarb became more involved in agriculture in 1990, he realized the root of all of our food systems and our whole agrarian society is based in seed and the work of seed breeders for many thousands of years, basically since the dawning of civilization.

And yes, commercial seed may be slightly cheaper, but at what cost? Is less cost worth sacrificing quality, quantity and time spent growing a seed from 45 to 120 days? You want the best seed for your needs and conditions, and you want a selection of the highest-quality organic seed to get the most out of that crop.

When you throw out the fabrications and formulas and simply reconnect to seed and soil by planting a handful of simple, wholesome seeds in your garden to grow something you eat, the crop you reap tomorrow is a windfall on the investment—and you live a better life too.

Chapter 3

COMMUNITY

The Modern Victory Garden

It should not be a privilege to eat good food. The idea of taking eating back into the hands of the people, by the people, for the people in a way that benefits the air, water, soil, community and commerce is rewarding. So we have planted a victory garden and invited the community to dig, plant and eat! FarmOn! at Empire Farm is the FarmOn! Foundation headquarters on a 217-acre Ag-Entrepreneurial working farm and community center with several local farmers contributing to youth education programming to teach students the business of food for viable livelihoods in agriculture. When Alice Waters hosted the ribbon cutting in September 2014, she exclaimed, "Happy farming!" You are invited to the farm to experience where your food comes from and change the way you eat.

Victory gardens originated during the First World War, when they were planted at homes and public parks in America and Europe by people who wanted to show their support for the war effort and reduce pressure on the public food supply. These gardens are more relevant today than ever before, as the Great Recession and increased public concern about eating a healthy and sustainable diet make us all ask if growing our own food, at home or, for city dwellers, in a communal green space, will make organic, truly local food finally available to all. Wouldn't that be a true victory!

The fact is, growing your own food (from organic seeds as explained in Chapter 2) is the cheapest, most reliable and, arguably, the most

Community victory garden at Columbia County's FarmOn! at Empire Farm, Copake, New York. *Courtesy Tessa Edick.*

fun way to step into the locavore lifestyle and impact consumption every day by eating one local ingredient each meal. It's powerful, empowering and supports the localized food movement with one simple step: planting.

MIT professor and FarmOn! Foundation board member Chris Weaver explained:

> *Our localized food movement is equivalent to the concept that spawned the distributed ethernet/Internet. In the 1960–70s, the U.S. military depended upon a limited number of large computer centers to provide critical services. The military was concerned that the Russians would bomb those centers and knock out military communications. A novel security alternative was conceived that would limit any one node to attack by creating a far more distributed network of many nodes—composed of smaller computers—thereby minimizing the loss of any one node to the network as a whole. Furthermore, the*

nodes were made intelligent such that if certain nodes went offline, the other nodes would sense this and automatically re-route around those failed nodes to maintain the integrity of the overall system. If you apply that same logic, what the locavore movement is doing follows a similar path for food distribution and supply integrity. By removing near total dependence upon a limited number of centralized distribution centers, the movement seeks to establish nodal "fail-safes" within the food distribution infrastructure. Not only would this localized food distribution save fuel and prevent widespread (accidental) distribution of contaminated foodstuffs—as we have seen far too often of late—but it would help revive the importance of local agricultural economies and reestablish the significance of the local farmer(s) within their immediate community.

At the same time the movement effectively "hardens" the food safety net for the populace, it (re)establishes clear need and importance of local agriculture. People become connected to the land again. They respect and value their farming brethren and the important service they provide to the community. And the economics of farming begin to reestablish themselves as a benefit to the local community—not some unnamed centralized Agricombine behemoth that cares nothing

Organic locally grown and harvested grain. *Courtesy Hillrock Estate Distillery.*

for the local community but is a business designed to feed only itself—often at the expense of the very populace it is supposed to serve. The system has become so broken that people have forgotten how agriculture used to exist—and how it worked for thousands of years—right up until the advent of mass transportation and refrigerated transport.

But just because you can have strawberries in December does not necessarily mean you *should* have strawberries in December. Humans understood that certain things grew at certain times in certain places and were OK with that for millennia. And what they could not know is that by accepting nature, they allowed nature to revitalize itself (i.e. green manures) during the time a natural period of "rest" was provided and not waste precious resources (such as oil) in the process—saving carbon footprint, saving unnecessary road use, saving local agriculture and saving local economies and communities in the process. Big business has proven time and time again that at a critical tipping point, it exists only to serve itself—not the populace it is supposed to feed and "protect."

So, the movement exists on many levels and for many purposes. It is not a political movement or a liberal movement. It is a movement "by the People and for the People" that seeks to underscore much of what made this country great—by reexamining some of the things that we have forgotten and recognizing the value of what these things brought to our grandparents and the generations before them—and wanting to restore that sort of benefit.

In 1995, there were eighty-one independent organic processing companies in America; ten years later, other companies had acquired all but fifteen of them, according to Philip H. Howard, a researcher and scholar at Michigan State University, who has created a widely referenced infographic entitled "Who Owns Organic?"

The chart was originally published in 2003, and it has been updated since to show the chain of acquisitions and alliances of companies that are commonly cast as the "bad guys" of the food world with the organic "good guys." (For example, Coca-Cola bought a 10 percent stake in Green Mountain Coffee for $1.25 billion; WhiteWave acquired Earthbound Farm for $600 million; General Mills owns Muir Glen, LaraBar and Cascadian Farm; J.M. Smucker owns Santa Cruz Organic and R.W. Knudsen; and the list goes on…and on.)

The Cornucopia Institute published an update of the chart this year. Download it at www.cornucopia.org.

Since previously iconic organic brands are now being run by the companies responsible for producing manufactured, over-processed food that has been linked to an endless number of health woes in children and adults, how long before the beloved organic companies start churning out sugar- and chemical-laden junk?

To slightly misquote Shakespeare, "Beware…the green-ey'd monster, which doth mock the meat it feeds on…"

I'm not suggesting that you boycott any form of organics, but if you are ready to dig in and get dirty in a patch of petrochemical-free soil, together we can revitalize the victory garden.

If you're new (or, like me, returning to gardening after many years), start small.

Top: Freshly picked strawberries in June ready for ABCSA pick up. Join a CSA today! *Courtesy Tessa Edick.*

Right: Early pear on a fruit tree in the rolling hills of Fix Brothers Fruit Farm. *Courtesy Cayla Zahoran.*

- PLANT. Thirty-two square feet can yield a lot of produce in a four- by eight-foot front yard garden box. They're easy to make, fill, plant, water, weed and manage. Weed control, pest control and disease control become a lot easier to deal with if you can hop, skip and jump across your victory garden. If you're really tentative, you can even just plant one pot of fresh herbs and keep it in your kitchen window!
- COMPOST. Start a compost pile in your own backyard. Not only will it cut down on garbage and control bad odor, it will also provide an amazing foundation for your garden bed. (Whole Foods has a simple and helpful how-to guide on its blog: www.wholefoodsmarket.com/blog/home-composting-basics.) Once you've got a nice pile, spread a thin layer of it over your soil beds a few times a year.
- SOIL. Create healthy soil by sourcing chemical-free options. Organic matter decomposition serves two functions for the microorganisms: providing energy for growth and supplying carbon for the formation of new cells. Soil organic matter is composed of the "living" microorganisms, "dead" fresh residues and the "very dead" humus fractions. Adding compost is key, as is creating walking paths between plantings so you can access without smothering sprouting seeds.
- MULCH. It's key. Keep it organic when possible.
- WATER. Plant roots need a deep soak on a regular basis to really sink into the soil—otherwise, they just hover near the topsoil.
- LOCAL. Make sure you plant native varietals in your growing zone (start with organic, regional seeds, as I mentioned in Chapter 2). Native plants thrive in your own region because that's where nature designed them to grow for thousands of years. Let's not reinvent the wheel.
- INTEGRATE. Show it off to your friends, and spend time in your garden! Add cucumber to your water, eat a fresh tomato off the vine or personalize Mason jars as saltshakers so your family and friends cheer on your victory garden too. It will center you for the day and make you feel even more connected to the land that feeds you.

Need more inspiration and guidance? My favorite go-tos are www.motherearthnews.com, www.organicgardening.com and http://smallfarms.

Get your garden going! FarmOn! *Courtesy Cayla Zahoran.*

cornell.edu. They give you actionable tips and guides, projects and resources minus corporate sponsors with Big Food, Big Agriculture and Big Seed influential agendas.

VARIETALS

There are certain fruits and veggies that just do better in our neck of the woods. You'll enjoy a heartier harvest and a higher-quality product, and they have built-in disease and insect resistance. And every month, there is something to do. Here's a rough calendar of planting to-dos, or refer to the NYS Pride website for Department of Ag & Markets at http://www.agriculture.ny.gov/f2s/documents/HarvestChart.pdf.

JANUARY/ & FEBRUARY: Let the land rest.

MARCH: Sow broccoli, Brussels sprouts, cabbage, eggplant, leeks, onions, peas and pepper indoors.

APRIL: Sow Brussels sprouts, cabbage, cauliflower, leeks, lettuce/ greens, radish, spinach and turnips outdoors.

MAY: Sow beets, broccoli, cabbage, cauliflower, corn, kale, leeks, lettuce/greens, peas, peppers, potatoes, pumpkin, spinach, squash, Swiss chard, tomatoes, turnips and watermelon outdoors.

JUNE: Sow beans, beets, carrots, celery, corn, lettuce/greens, melon, parsnips, peas, pumpkin, radishes, spinach, squash, sweet potatoes and watermelon outdoors. Harvest broccoli, eggplant, lettuce/greens, onions, peas, peppers, spinach and tomatoes.

JULY: Harvest beans, broccoli, cabbage, cauliflower, eggplant, kale, lettuce/greens, peas, peppers, radishes, spinach, squash, Swiss chard and tomatoes.

AUGUST: Harvest beans, beets, broccoli, cabbage, carrots, cauliflower, celery, corn, eggplant, kale, lettuce/greens, melon, peas, peppers, potatoes, radishes, spinach, squash, Swiss chard, tomatoes and watermelon.

SEPTEMBER: Harvest beans, beets, broccoli, cabbage, carrots, cauliflower, celery, corn, eggplant, kale, lettuce, melon, peas, peppers, potatoes, radishes, spinach, squash, Swiss chard, tomatoes and watermelon.

OCTOBER: Harvest beets, carrots, celery, kale, leeks, lettuce/greens, potatoes, pumpkin, Swiss chard and turnips.

NOVEMBER: Harvest carrots, celery, kale, parsnips, potatoes, pumpkin, Swiss chard and turnips.

DECEMBER: Harvest Brussels sprouts, kale, leeks, parsnips and Swiss chard.

Right: Just picked fresh corn at Holmquest Farm in Hudson, New York. *Courtesy Cayla Zahoran.*

Opposite: Harvest Hall at the Columbia County Fair in Chatham, New York, features the best of local produce and celebrates the farmers' harvest and hard work with award ribbons each August. *Courtesy Tessa Edick.*

PRESERVING AND CONSERVING

Skip food with labels, and "can all you can," as a famous victory garden promotional ad suggested. Preserving and conserving just-harvested vegetables and fruits (even if they're fresh from the market) is one of the most sustainable and most delicious ways to make your own food. And your friends will swoon. Earl grey peaches, anyone? Canning recipes and how-to guides are in abundance online, but as with all things in life, choose the path of least resistance for success:

- Prep. Gather your (sterilized!) supplies: water bath canner, fun and cool canning or Mason jars, lids, rings, a jar lifter, a funnel, towels and pot holders, pots and bowls, spoons, knives, cutting board and premade ingredients. (My favorite sources for pickle, jam and other canned fruit and veggie recipes are seriouseats.com and the kitchn.com. Simple, delicious, fab.)

Fresh blueberries, Sungold tomatoes and strawberries at the height of summer beg you to try your hand at canning, pickling and preserving! *Courtesy Common Hands Organic Farm.*

- Fill up water bath canner two-thirds full for pint jars, half full for quart jars, with hot water but not quite boiling.
- For pourable canned goods, like jams, ladle the mixture through a funnel into the jars and leave about a finger's worth of space between the jar rim and the food. For non-pourables, place food in the jar with a large spoon and then pour any hot liquid through funnel over the foods, leaving about a finger's worth of space again. To release air bubbles, run a plastic knife around the lip of the jar, and wipe the jar rims clean. Place seals and rings on jars and tighten the lids.
- Using your jar lifter, place the jars in the rack in the water bath canner (which prevents them from being jostled and breaking). Make sure water can flow freely between jars and that they are covered by water (about two inches should suffice).
- Cover, crank up the heat and allow it to boil. Time the boil according to your recipes.
- Turn off the heat and carefully remove jars from the water bath with your jar lifter. Set on a towel to cool. Check the seals (if you press the lid and it sucks down and doesn't pop back, it's solid). If any didn't "take," put them in the fridge and eat within a week.
- Label, date and store in a cool and dry environment.

ITALIAN PLUM JAM (SERVES 6)
Recipe by Paula Colarusso (@thefoodishgirl)

Italian plums, also known as prune plums, are succulent and rich. A few ingredients and a little time and heat create a decadent fruit spread. From large orchards to backyard gardens, Italian immigrants of the Hudson Valley always have a few prune plum trees to make jam, brandy and wine.

INGREDIENTS:
5 pounds Italian plums
4 cups of sugar
1 tablespoon of whole fennel seed

METHOD

The night before: wash, pit and chop the plums.

(No need to chop finely; they cook down, and you want bits of plum on your spoon when this is done.)

Put the plums in a bowl with the sugar, cover and place in the refrigerator overnight. Prune plums are very fleshy and need this rest with the sugar to macerate and create juice.

The next day: In a cast-iron or similar heavy-bottom pot, dump in the plums and sugar and cook over medium heat stirring often.

Place the fennel in a square of cheesecloth or in a tea strainer and add to the pot. Do not add the seeds directly to the jam.

After about 20 minutes, taste.

Each batch might be a little different, and usually no more than 4 cups of sugar is needed.

You can cook to a temperature of 220 degrees, which would offer a firmer set, or a little under that temperature if you want the fruit to be spreadable, which is preferred.

Plums are a high-acid fruit, and while sugar is a preservative, you can safely lower the sugar in this recipe without compromising safety.

Your plum mixture should coat the back of a spoon like a veil of purple syrup when it is ready to process.

Ladle hot jam into sterilized jars, wipe the edge clean and cap with sterilized lids and ring.

Process in a boiling water bath canner for 10 minutes per standard canning practice.

ADD-IN OPTIONS:

Delicious with local cheese.

Stir into Greek yogurt.

Top your favorite scone or corn muffin with a little local butter and jam.

Add to your favorite brownie recipe.

Chapter 4

COMMERCE

Impact with Food Choices

What is natural, what is homemade, what is hand-crafted? Natural food means unprocessed, nonchemical food that is good for you, the environment and farmers, right? Not necessarily. Before making an investment in your food choices in a supermarket for something labeled "natural" or "farm-washed," think about what you're spending your money on, the nutrition provided and who benefits from your food dollars.

The global market for organic goods reached about $64 billion in 2012, and in the United States, it's expected to grow at a compound annual growth rate of 14 percent from 2013 to 2018, according to Research and Markets. And that's nothing: in the last twenty-four years, there has been a 3,400 percent increase in organic sales, making it the fastest-growing consumer food and lifestyle "trend" in modern history, Food Safety News reports.

So this means that consumers are demanding better food and becoming aware of organic benefits, the vital role of the family farm and how the government is selling and marketing a certified seal of approval "USDA organic," but what does this really mean? Are we being sold a side of hype alongside our kale that was once a lowly garnish?

Unfortunately, what we'd like to believe in as our God-given right to good, honest food for all and implementing common sense with regard to labeling and consumption is not at all applicable when it comes to shopping for food attached to a convenient lifestyle we swapped for nutrition. "Natural" should mean from nature, but some of the products with this label are seemingly anything but.

Roadside farm stand at Sir William Farm in Craryville, New York. *Courtesy Stephen Mack.*

Feeding time at Pigasso Farms with Heather Kitchen and her pig posse in Copake, New York. *Courtesy Cayla Zahoran.*

In America, there is no official USDA-sanctioned definition of "natural," so foods with chemicals, artificial ingredients and all kinds of things that are pretty much the opposite of certified farm-fresh appear on ingredient lists as allowed.

When it comes to labels, we're all confused. Here's what organic versus natural is, according to the USDA:

- Toxic persistent pesticides: organic (not allowed) versus natural (allowed)
- Genetically modified organisms (GMOs): organic (not allowed) versus natural (allowed)
- Antibiotics: organic (not allowed) versus natural (allowed)
- Growth hormones: organic (not allowed) versus natural (allowed)
- Sludge and irradiation: organic (not allowed) versus natural (allowed)
- Animal welfare requirements: organic (yes) versus natural (no)
- Cows required to be on pasture for pasture season: organic (yes) versus natural (no)
- Lower levels of environmental pollution: organic (yes) versus natural (not necessarily)
- Audit trail from farm to table: organic (yes) versus natural (no)
- Certification required, including inspections: organic (yes) versus natural (no)
- Legal restrictions on allowable materials: organic (yes) versus natural (no)

I don't know about you, but pesticides, GMOs, antibiotics, growth hormones, sludge and irradiation don't sound too "natural" to me.

Deciphering the labels is just the tip of the iceberg. Organic food (once known simply as "food") isn't at all what it used to be, though most consumers aren't aware of the changes that have completely re-created the industry. Corporate consolidation of the organic sector of the food world has been largely hidden from consumers.

Philip H. Howard, a researcher and scholar at Michigan State University, created an extremely extensive chart (available online at https://www.msu.edu/~howardp/organicindustry.html) that clearly illustrates how the leaders of the snack world have started gobbling up the biggest presences in the organic, sustainable, free-trade world. I can't re-create it here (hello, copyright infringement!), and it truly does have to be seen to be believed, but here are just a few key examples of what Howard has illustrated: junk-food kingpin General Mills owns organic paragons LaraBar, Muir Glen and Cascadian Farm, and Coca-Cola owns Honest Tea and Odwalla.

Black Angus beef is America's brand of choice—choose local to make your meat matter. *Courtesy Cayla Zahoran.*

And since he most recently updated his chart, even more acquisitions have occurred: in April 2014, Post acquired Michael Foods for $2.45 billion, Hillshire Brands gobbled up Van's Natural Foods for $165 million, Treehouse Foods acquired Protenergy Natural Foods for $150 million, Hain Celestial acquired Rudi's Organic Bakery for $61 million and General Mills acquired Annie's Homegrown. Are we destined for all of our food to be franken-ingredient laden?

OK, you say. If I can't build my diet around food labeled "organic" or "natural," what can I eat? Good question.

Most of us still build our eating plans primarily around dairy, eggs, meat, fish, grains and vegetables. And we know we are being tricked. Who is fooling us most? What can we eat? Which of these foods actually contain nutrition, and where do they come from? What are we actually eating? While many of the label claims imply a concern for our health and wellness and the plants and animals we eat, the reality is really very different.

Crack both a store-bought egg and this morning's freshly laid egg into a white bowl. Which would you choose for breakfast?

Poison for profits is the name of the game, and navigating the world of food is indeed tricky. Labels lie. Certifications aren't credible, and eating organic is expensive and not widely available for so many reasons. Worse yet, many organic products are sourced from tens of thousands of miles away or China and aren't even an option for all for so many reasons. It's all gotten too complicated.

We'd all like to believe in the idea of organic for the land, for the air, for the water, for the future of farming and the wellness of our children and theirs, but is eating organic really even possible? The majority of our food comes from uncertified non-organic sources, and the mono-crops' use of petrochemicals makes us question contamination and whether an organic label really ensures organically grown food, taking into consideration factors like wind, bees, butterflies and cross-pollination. It makes you ask questions, which is essential on the path to finding better food ways.

The Environmental Working Group releases a list of the most pesticide- and toxin-contaminated produce every year, a powerful tool to have in your arsenal (http://www.ewg.org/foodnews/summary.php).

So, what can we eat?

I say meet your farmer. I say get connected to seed and soil. Source responsibly and research to have a better understanding of terroir and its composition, based on where you live, best practices in making food, how the animals live and their welfare and who is making your food.

Remember, you are what you (and they) eat. That's why eating local is a "best practice."

When you meet someone feeding you, you trust them to not poison you just for a few extra bucks. So at any point that you can, direct your food dollars to the family farms for food shopping. Make a list, make it a priority and this lifestyle will change the way you eat.

The Hudson Valley has been celebrated as a breadbasket, thanks to the plethora of family farms and the rich gastronomic region that makes it easy to eat local. Not everyone is as lucky as I am to source 70 percent of their food within a five-mile radius of their home, which optimizes nutrition and therefore your health, as well as your appetite, so obesity becomes a non-issue. When you eat in season, you get what you need from your food, so you look and feel great. You return to the kitchen, cooking and preserving what you might want off-season, and learn to love strawberries in June and preserve them so when the craving comes in January, you're still eating right.

People say that's inconvenient. People say that's expensive. But what is good health worth to you? And what's good taste? And what's knowing your food choices impact the family farm, commerce and

Heather Kitchen of Pigasso Farms raises animals with good food, love and care. *Courtesy Cayla Zahoran.*

Peppers picked from the field at Holmquest Farm pack a punch and crunch we all crave. *Courtesy Cayla Zahoran.*

your local community worth? I say it's just worth it to invest in the locavore lifestyle as a food source, as a family activity, as a right to consumption.

Take baby steps into better food. It starts with dairy. Grilled cheese for the kids, milk in your coffee and yogurt offer the easiest route to choose local sources. A quality glass of milk comes from comfortable, happy cows kept clean and often found grazing on the Hudson Valley hillside to offer us a low somatic cell count (the lower the count, the cleaner the dairy) and highly nutritious milk that tastes better than almost anything (I love raw and goat milk too!). People are so disconnected to nutritious whole milk, more worried about fat content and the organic word than the nutrition so vital to optimizing health. The plethora of commodity-driven grocery store brands (yes, even the organic ones) have made us all forget what the real stuff is like—from a cow in a field, eating, living and breathing to feed you well.

ALL MILK IS NOT CREATED EQUAL

The Hudson Valley Fresh dairy cooperative's tagline is "all milk is not created equal," and Tollgate Farm is the reason why. One of the family farms responsible for providing HVF with milk, it has won thirty-four prestigious top-quality milk awards, including the 2013 national quality award, a testament to Jim and Karen Davenport's dedication to premium-quality milk production at Tollgate Farm in Ancramdale. Congratulations! Thank you for caring so much about the cows and feeding us lucky folks in the Northeast.

The sign at the gate of Tollgate Farm reads "A Dairy of Distinction," and it lives up to its name on 140 acres with views of the Hudson Valley that you dream of. When I walked in, Farmer Jim was putting fresh hay into the calves' houses "so they will be warmer and more comfortable in the cold," he explained.

Jim and his wife, Karen, have smiles on their faces every time I see them. They love what they do, and I love them for it. Milk from their farm isn't only healthier because it is more nutritious, but it also helps support commerce in our community and in the long run saves you money. Taking an active role in helping farmers like the Davenports sustain profitability in agriculture is vital. Clearly, they are on the right path, growing their business locally in a cooperative farmer-owned model that we can all support by simply making informed choices when food shopping.

Speaking of a dream, clean, comfortable and happy is the cow motto at Tollgate Farm because when cows aren't stressed, they produce better milk. Jim and Karen Davenport have been raising purebred cows under the Tollgate name since 1986, growing enough grass for hay and corn for silage to feed their current herd of sixty-four milking cows and eighty young stock at the farm in Ancramdale.

Each of their cows is registered and named properly with the "Tollgate Vu" prefix that is indeed enviable. These purebred cows (90 percent Holstein and 10 percent Ayrshire) are free-range, grass-fed and milked in the original tie-stall barn that came with the property. Jim likes the amount of room and ventilation in the barn. Each cow has its own fitted stall, determined by its length and width.

Jim is fastidious about maintaining a clean farm. "When I come in the barn and turn on the light in the early morning and I see all the cows are clean and all comfortable, well, that is what makes it worth doing. If I'm going to be in the barn at 3:00 a.m. on Christmas

Columbia County Dairy Princess and Ambassadors promote local dairy to consumers. *Courtesy Emmanuel Dziuk.*

morning, I want to work with cows that are attractive and nice to work with," Jim explains joyfully.

The daily to-do list is plentiful at the farm, and Jim must be bionic, because his sleep-deprived day, every day, starts at 2:40 a.m. when the alarm sounds. He's in the barn by 2:55 a.m. to sanitize the milk; feed the cows; and clean, scrape and pamper them and their stalls so it is spotless where they lay down. At 3:45 a.m., milking starts, and fresh milk is cooled fast in a stainless-steel tank, the key to quality.

It is then shipped via a double-insulated truck to Agrimark and Boice Brothers Dairy for processing and pasteurization to produce Hudson Valley Fresh milk, cream and yogurt that is never co-mingled with other milk outside of HVF; Cabot cheese; McAdams cheese; and many other products from cream that is separated during production to make sour cream, ice cream and different types of dairy products for specialty food makers.

Even Whole Foods Market bakeries demand Tollgate Farm milk because the Davenports consistently deliver premium dairy products, and you can taste the difference.

Agrimark is a farmer-owned cooperative with 1,200 local farmer members who feed their herds, their families and communities honestly.

It's important to know where our milk comes from and who makes it. And we should be willing to pay for that.

We will pay $2.00 for a twenty-ounce bottle of water—the cheapest ingredient used to nourish cows—but balk at $3.50 for sixty-four ounces of nutritious milk, opting for bigger corporate brands that are cheaper, don't taste as good and are far less nutritious. "It makes farmers bonkers," Jim exclaimed. It makes me think about food choices I make three times a day.

As an agriculturalist and fifth-generation dairy farmer, Davenport's farming journey from Massachusetts to Connecticut to New York gives him insight and understanding about high standards necessary to make milk we love—and that will win awards.

Cows love to eat—they live to eat, in fact—and "by feeding them properly we make better milk," Jim told me. And you cannot imagine the science involved in feeding them "for spot-on nutrition" essential to make his economic model work and for the cows to yield ten gallons of milk each day—two milkings every twelve hours—free from added hormones like rBst or antibiotics.

"If we don't feed to genetic potential, we couldn't afford to stay in business," Jim explained to me in micro detail, stressing the importance of what is fed to cows. Their mix of forage, starch and grass is essential to yield protein-rich milk full of nutrition so important for human consumption. It's beyond my comprehension, but clearly, Jim's talent is a passion and sure makes Hudson Valley Fresh milk a staple in my fridge!

All this has been going on for generations in Jim's family—his grandfather Preston Davenport carried on the dairy farming tradition since 1939 and was Jim's inspiration, having started Tollgate Farm in Litchfield, Connecticut, named for the tollgate at the end of the road. The name stuck, and after Jim graduated from college, he bought his own Holstein cow and carried on the family legacy in dairy farming. Jim initially studied engineering but wanted to be in agriculture to humor his mother.

Jim and Karen met in college, both graduating from the University of Connecticut with bachelor's degrees in animal science in 1983. Karen went on to earn her master's degree in animal science and an education certificate and is currently the department head of the agricultural education program at Housatonic Valley Regional High School in the northwest corner of Connecticut, where she teaches a diverse curriculum, including biotechnology, animal science and life skills and chairs Future Farmers of America. Karen makes learning about farming fun and

Local dairy means artisanal cheeses of many types for family farms. *Courtesy Cayla Zahoran.*

engaging for kids—she is a real-live dairy queen and heads up all the farm tours and promotion of the products Tollgate produces.

The same practices on the farm are applied in the Davenport home, as their two daughters, Kristen and Lauren Davenport, are now top performers too. "There was no spare time to get into trouble," Jim said. "Raising kids in ag and having a gifted teacher like my wife gives a foundation for success that money cannot buy."

Working four thousand hours a year, Jim has "quality milk production down to a science—but I'm not getting rich from it," he says with a laugh as he looks around the barn smiling while Karen sweeps the corridor. Their eldest in the herd, called Sapphire, seems to be smiling too. She continues to produce milk at fourteen years of age.

These cows are part of the Davenport family at Tollgate Farm, and the atmosphere in the barn is relaxed and cheerful because of the hospitality Jim and Karen show the cows—treatment many cows don't receive. I see why cows are easily loved, and today I appreciated their hard work too.

Take your own milk taste-test challenge at home with Hudson Valley Fresh, and trust me—you will seek out quality milk from local family farms that use best practices to farm responsibly.

Dollar for dollar, the investment in honest, local food and eating responsibly will result in a better taste experience. It's guaranteed to improve your health, revitalize your community, stimulate economic development and ensure the family farm is sustainable, all a result of spending your food dollars wisely.

Small-scale farms with gross incomes of $100,000 or less made close to 95 percent of farm-related expenditures in their own community, contributing to the overall health of their towns, according to a study from the University of Minnesota. Factory farms, on the other hand, often spend their equipment and supply dollars at the big corporations that pledge to purchase their products. Less than 20 percent of farm-related expenditures are made locally, according to the same study from the University of Minnesota. Smaller farms are more rooted in their communities, literally and figuratively.

It's time to start looking at companies selling food that aren't holding themselves to a higher standard of best practice, transparency and quality and instead commit to buying your food from farms that are proud to produce nutrient-dense food for you to eat. Reconnect your children to resilient agriculture and better health to create strong rural-to-urban marketplaces in our communities. Personally, it's my mission.

In this country, within the next ten to twenty years, the majority of our American farmers (average age fifty-eight) will age out of their profession, and millions of acres of farmland will change hands and need new farmers. Who is going to feed us?

Of course, food is personal—it is both a choice and a lifestyle. The reward for practicing consumption of organically grown (not only labeled) food from the farm made by people you can trust to implement best practice for production is a gift from the food gods—otherwise known as farmers.

At Hawthorne Valley Farm in Ghent, New York, it's a wonderful journey into the land of biodynamic organic food, sustainable practice and preservation of food justice—our human right to eat fresh local food and be well.

"Good, healthy food cannot be a privilege; it has to become a basic human right," said Steffen Schneider, director of farm operations at Hawthorne Valley Farm.

Food from Hawthorne Valley Farm is exclusively GMO free, biodynamic, certified organic, pasture raised and—as luck would have it—affordable.

Three little local pigs kissing in the Hudson Valley. *Courtesy Cayla Zahoran.*

People are just learning what these practices mean—and learning that just stating "organic" on the package doesn't mean a product isn't worth investigation. "It's not obvious in people's minds that organic is supposed to mean GMO-free—it's becoming another buzzword," Lauren Wolff explained as Hawthorne Valley Farm marketer.

The signage at Hawthorne Valley Farm reads clearly, "Our produce is biodynamic and organic," but what does that mean?

If you read on, it says: "Biodynamic farmers will strive to 'root the farm in the whole household of nature' and will include the conscious practice of working with the rhythmical influences of the 'cosmos' and strictly avoid synthetic inputs (fertilizers, pesticides and hormones) on a Biodynamic farm."

As a biodynamic farm, it also means that the amount of land determines the number of cows that can be fed. (In other words, X number of acres can only feed X number of cows. The size of the farm naturally limits the number of cows it can biodynamically sustain.) In this case, four hundred acres allows a herd of sixty Brown Swiss dairy cows to supply organic milk to the farmstead creamery.

It's all about best practice, really—transparency and quality—to ensure your food is packed with nutrition from land that is respected; water that is clean; and animals that are well cared for, comfortable and happy to create a marketplace that brings education and economic development to the community. Why doesn't every package of food you eat say that?

This working farm also includes ten acres of vegetables that support a three-hundred-member CSA and a two-acre Corner Garden, which provides produce for the farm store, open year round. Twenty pigs eat the "waste" whey from the creamery, and forty chickens provide eggs for the learning programs on the farm. "When a three-year-old is nose to nose with a baby pig, it's so funny and personable. It's a way to trust food and sustainable practices," director of marketing Karen Preuss shared.

So this four-hundred-acre farm and farmstead creamery work within the limits of the biodynamic scale. There is a ration of land available with nutrient-dense pasture/grass to graze for the feeding of these sixty lovable and child-friendly Brown Swiss milking cows, all of which produce milk used in the creamery and sold in many forms, including raw milk.

Raw milk is an unpasteurized, nutrient-rich dairy product that is both coveted and feared because the industry has taught us that milk should last a long time in your fridge and pasteurization is mandatory—unless bought directly on the farm.

If you have never tried it, try it! Hawthorne Valley Farm raw milk is available in the farm store, and the taste is VERY different from the pasteurized version many would call "white water" in organic terms. It has a one-week expiration date and is what I call a real energy drink!

In New York State, you can only buy raw milk direct on a farm property where it is handled correctly to mitigate any risk with cleanliness and testing. It is the most nutritious milk you can buy. People come from far away and freeze it, so if you do visit the farm to buy raw milk, call ahead and bring a cooler to transport it correctly to your fridge.

Neighboring farms lend milk from Jersey cows. HVF varies the milk sources to diversify the cheese types made at the on-site farmstead creamery. There are several types of cheeses made at the creamery. Alpine is made only in the summer and takes eight months to mature. It is crafted from raw milk and aged sixty days (by law) on property in the cheese caves. It comes in caraway and classic flavor—all of it, of course, sourced from biodynamic organic milk from the Brown Swiss beauties on property.

"Cheddaring" is a cheese-making process from Cheddar, England, and salt is everything to cheese in this labor-intensive process. America's favorite cheese—cheddar—is made at Hawthorne Valley Farm in the

Farmstead creameries are Hudson Valley treasures worth trying! *Courtesy Cayla Zahoran.*

"clothbound" tradition (wrapped in cloth and aged in a cave or cellar at certain temperatures with specifically calibrated levels of humidity), a version that will make you swear off cheap imitations. The farm store offers many varieties, all aged in caves for year-round eating, based on the philosophy that it is a working farm committed to meeting its community food needs and offering a variety of products in small batches to feed us well.

There is also raclette in winter (nutty, rich and perfect for fondue) and a summer havarti (mild and of the famed "Edamer" ilk), as well as a May Hill "camembert-style" chèvre. Since the cows graze on May Hill, the name reflects the terroir. Euro-style yogurt is a favorite at Hawthorne Valley Farm in plain, maple vanilla, strawberry or lemon flavors, all made from whole milk from the same Brown Swiss cows. There also "Bianca" cheese spreads, much like soft goat-style cheeses. These are great for salads in many flavors: classic, smoked, olive, chive or herb.

Because Hawthorne Valley Farm is so invested in educating children and members of the public who want to learn more about agriculture, it offers a variety of programs that appeal to and consistently draw a variety of people from Columbia County to Manhattan and beyond. Hawthorne Valley has a progressive school, a variety of productive and ornamental gardens, a kraut cellar (fermented foods!), a sprawling farm store, a dairy herd and a creamery, so it not only serves the immediate educational and nutritional needs of the community, but it also caters to our desire to invest in a responsible, sustainable agricultural system—and eat amazing kimchi to boot.

Visitors go and return, again and again, to Hawthorne Valley Farm, not just for its addictive barn-to-belly treats but also because it teaches them to slow down, understand the process of getting those treats from soil to our bodies, consider the implications of the process and connect not just to the food they eat but also to the people and animals that produce it.

As Executive Director Martin Ping explains, "We rush toward progress, but progress toward what? We are in the business of growing food. Why are we on the defensive? We should trust cows more than chemists."

Hawthorne Valley Farm is funded by a diverse mix of revenue streams and has a social mission with a pedagogical entrepreneurial instruction. Funding comes from the store, grants, the on-site summer camp and educational programming. From pre-kindergarten to twelfth grade, this

Waldorf/Steiner school educated 275 students in the 2013–14 school year. (The Waldorf/Steiner educational system is based on the educational philosophy of Rudolf Steiner, the founder of anthroposophy. There are Waldorf/Steiner schools throughout the world, and the system is well regarded by many for its holistic, humanistic approach to learning.) It is actively growing the high school boarding program with an international program hosting students from as far as Afghanistan, China and Israel to study at the Hawthorne Valley Waldorf School, a member of Hawthorne Valley Association, a diverse not-for-profit committed to social and cultural renewal through the integration of education, agriculture and the arts.

It started with the philosophy of Rudolf Steiner, who said, "There is no realm of human life that is not affected by agriculture." According to the Hawthorne Valley Farm website, this led to

> *a group of pioneering educators, farmers, and artisans to take a courageous step into an uncertain future when they purchased the Curtis Vincent Farm in Harlemville, New York. This deed was the culmination of a seven-year process that began in response to experiencing firsthand the immediate challenges of the loss of small family farms and the threat to childhood development posed by an increasingly materialistic and mechanistic prevailing worldview. The idea was to buy a farm and offer children from urban centers a hands-on experience of what it means to be stewards of the land. In 1972, the first class of visiting students from the Rudolf Steiner School in New York City drove to Harlemville, New York, and renovated the farmhouse into a bunkhouse where the Visiting Students Program continues today, having hosted more than thirteen thousand children at this writing. Many local neighbors involved at the beginning still refer to this initiative and Hawthorne Valley as "the farm school."*

These founding impulses—sensitive land stewardship, healthy child development in connection with nature and provision of nutritious food—continue to guide the work at Hawthorne Valley Farm today. The five full-time farmers and eight apprentices work hard year-round to tend the plants and animals in addition to monitoring the health of the whole farm and connecting people to nature and our food supply. As Steffen Schneider, director of farm operations, pointed out in his "Agriculture 3.0" lecture, they do this with "a renewed awareness of the fundamental importance of a regenerative, resilient agriculture for our society as a whole."

It's a philosophy you can invest in—especially when it comes to good food and doing good, just by eating.

Chapter 5

SUSTAINABILITY

The Locavore Lifestyle

O ur dollars drive demand, and production follows our demand. When we demand sustainable food, we are doing part of our part, but what about cultivating a locavore life? It's not always possible or even desirable for that matter to ensure that every morsel we eat was plucked from a five-mile radius of our plate, but really, the only vital part of each day is feeding yourself, so making those food choices wisely is an investment. Do we only appreciate our good health once it is gone?

The Europeans have the right food philosophy! Everyone is always saying how good the food is in Italy, France and Spain, but honestly, it's not a big secret as to why: they shop every day for fresh food from the farm made without chemicals on quality soil from heritage and heirloom seeds that are treasured and cultivated with care, and the European Union even protects local and regional food production certification "designation of origin" and "geographical indication," known as DOP or IGP.

It was in the small villages of Europe, where farmers (who were honored by their countrymen like we honor chefs) fed their own leftovers to the animals that would feed them later on, that I first began to truly understand the passion for where our food comes from and what goes into making it.

In Europe, there are laws that safeguard and are designed to boost regional specialties. Gorgonzola, prosciutto di Parma, bacalao, truffles, Champagne and Barolo—the products are all familiar by name (and

taste), and the practices of making them have been passed down by generations. Though many of us know the products, many have no idea that the names also refer to the region in which they're produced and are protected like national treasures. We should consider this in the Hudson Valley, which is ripe with opportunity to heighten awareness and bring tourism to the region.

But many farmers and consumers, in recent years, are becoming aware of the shift in thinking when it involves farm to your plate. We want our food and our lives to reflect who we are, what values are important to us and pride in our place. We want to know where our food comes from, who raised it and how they raised it—but sacrificing convenience is the longer haul. We have started on a path at restaurants as they source farm to table. Now is the time to replicate that model—farm to *our* tables—and pack our cabinets, cupboards, fridges and freezers with food direct from the farm. Know your farmer. Know your food. Meet your farmer. Connect to what you eat and why it matters.

When consumers demand accountability, sustainability and responsibility from the companies they purchase products from, corporations comply. They care about profits. And the farmers, despite their reputation as simple country folk, are astute business people of character. They are ingenious and will grow what we demand. But we have to clearly tell them what we want to eat, not let corporations supply what they think we should eat to fill their pockets with money, supplying empty calories and an epidemic of sick, fat, sleepless, sexless and moody consumers.

Between 2010 and 2030, global demand is projected to grow by 33 percent for primary energy, 27 percent for food and 41 percent for water, according to the United Nations Environmental Programme Report titled "The Business Case for the Green Economy." There is a finite pool of food resources and many more mouths to feed, which leads to constraints imposed by space, time, environment and ignorance.

The day we all decided to use reusable bags and abandon their plastic counterparts with an eco-conscious consideration and common courtesy has to be replicated in the supermarket with our demand for better, nutrient-dense food. It's our obligation to health and wellness.

Opposite: Handpicked Rainier cherries. *Courtesy Cayla Zahoran.*

Grass-fed Black Angus roam the rolling hills at Sir William Farm in the Hudson Valley. *Courtesy Cayla Zahoran.*

FarmOn! is a movement—a shift in food rules and thinking—based on the idea to do what you can to opt out of processed food and feed yourself and your family better without sacrificing taste or polluting the environment.

Moving toward sustainability is a social challenge that makes us all reexamine our lifestyles, ethical consumerism, living conditions, work practices and technologies that impact our natural resources and food shed. The idea is to maintain and support development. As the Brundtland Commission of the United Nations on March 20, 1987, put it so clearly: "Sustainable development is development that meets the needs of the present without compromising the ability of future generations to meet their own needs."

Small changes make big impacts: if every day, once a day, 300 million people in the United States alone demand just one local ingredient in just one meal, what would happen?

I've shared intel on the importance of organic seed and planting your own victory garden and introduced you to a few Hudson Valley farmers, but this book at the very least is also about finding other ways to inspire real change in your food habits each and every day—one ingredient, one meal, one day at a time. Let's face it, maybe not everyone is ready to trade in their Louboutins for muddy muck-boots and not everyone has the space for a victory garden, but wouldn't planting a garden and meeting your farmers inspire real change in your food habits every day?

Wellness and nutrition make you look and feel great. If you shift your thinking into an understanding that your food MUST come from a farm, you can ask questions about food shopping to help you eat better from the produce manager, the butcher, the cheese monger or your grocer. With online delivery services, food subscriptions, CSA programs, farmers' markets and supermarkets making the switch to be informed and inform you, all you have to do is ask: where does this food come from?

FarmOn! Ed Kuester, Kristopher Kelly, Tessa Edick, Stephen Mack and Brian Schaefer converted to the locavore lifestyle. *Courtesy Ed Kuester.*

Most of you can be like me and source your food within miles of your home from farms that work hard to feed you well. You can ask where your food comes from and get involved with real food, better choices and responsible eating just by meeting your farmer and eating local.

Since 2008, there has been a 76 percent increase in the number of farmers' markets, according to the National Farmers' Market Directory. California is tops (764 markets), but New York (683 markets) is hot on its heels. All said, there are more than 8,200 farmers' markets across the country.

That isn't enough, but we're continuing to grow. Consumers want to shake the hands that feed them. There is indeed something so beautiful and right about seeing and touching and learning and eating straight from the source—never mind knowing that the money you give farmers in exchange for those fragrant, red-ripe strawberries is helping them directly, not going mostly to middlemen and distribution.

Direct sales from farmers to consumers are leading agricultural sales, according to the USDA. I often use a great online resource at www.localharvest.org when I'm in an unfamiliar town or state throughout America. It is dedicated to guiding people to community-based food systems. (For some of my favorite farmers' markets in the Hudson Valley, turn to my farmers' market guide on page 169.)

Not everyone is fully committed to shopping their local market every time they want a tomato, but look at your options—and your reward from going out of your way to eat locally sourced food. Instead of hitting up a big box store, get your feel-good foodie on with Field-Goods.com. Field-Goods.com is another Hudson Valley resource I use that is replicated in many regions outside the Hudson Valley. It makes it so easy to eat affordable, locally sourced quality food that provides honest claims about sourcing and practices used for farming to aggregate food from small farms and make a big impact in the business of food, making it convenient to shop.

Donna Williams started Field-Goods.com in 2011 after completing agriculture economic development consulting work for the Greene County Industrial Development Agency. Field-Goods was an inevitable outgrowth of this work, and she shared, "My belief is in the positive power of local food and small business, which explains my venture in consumer health, e-commerce and the natural food industry."

Williams went on, "Field-Goods' mission is to open the floodgates for the flow of fresh food from small farms to urban and suburban

A locally sourced, organically grown, farm-fresh heirloom tomato. *Courtesy Cayla Zahoran.*

communities. Our goals are to bring food of superior taste, freshness and nutritional value to a large audience and to offer an accessible and profitable channel for small producers to sell their product."

Here's how to get started with a better way to eat. It takes as few as five folks, depending on location and given the many drop spots. You select your pickup location. Sign up online to select food quantity. Manage how many weeks you want groceries from small farms. Arrange a pickup time for all. There's even a hold option for weeks you are away!

Quantity varies from "family size" to "small size" portions of local produce—vegetables and fruit—as well as add-ons of pasta, cheese and local loaves of fresh bread.

Subscriptions range from twenty to thirty dollars per week and are accompanied by a newsletter that tells subscribers what they'll be getting and what to do with it. Field-Goods.com delivers more than one hundred different varieties of produce and purchases its products "to-order" from farmers in the Hudson Valley.

This model allows the company to minimize inventory loss and offer its customers exceptional value. Another option for folks closer to New York City is a CSA program I "citified," meaning I made it city friendly

At the FarmOn! Hootenanny! fundraiser dinner, Chef Jean-Georges Vongerichten's ABC Kitchen recipe for simple bean salad makes any meal indulgent and fresh from the farm. *Courtesy Cayla Zahoran.*

for city dwellers who don't need twenty types of vegetables each week that they don't have room to store or time to cook. The program was curated exclusively for ABC Kitchen, the beloved farm-to-table gem of a restaurant off Union Square. ABC Kitchen in Manhattan is a favorite restaurant not only because it is fun, sexy and serving delicious, honest food but also because Chef Jean-Georges Vongerichten is passionately committed to offering the freshest organic and local ingredients possible.

The restaurant is located in the shopping mecca of ABC Carpet & Home on East Eighteenth Street near the Union Square Greenmarket. A culinary vision of Jean-Georges and Phil Suarez, it is committed to an organic, sustainable and meaningful practice in food. The locally sourced menu is always changing and engages the diner in cuisine that is stated in its mission to be "rooted in cultivating a safe relationship with the environment and our table."

ABC Kitchen's active commitment to local and organic produce, meat, fish and dairy sourced from humanely treated and pasture-fed animals honors the food chain and people who produce the food we love to eat and share at any table.

The chefs' relationships with farmers guarantee the freshest ingredients available for their recipes. Knowing your farmer is key. You know your food—how it was cared for, when it was harvested and who grew it. This is the secret to delicious goodness and food bursting with flavor, nutrition and satisfaction. This is the well-kept secret of a chef. Now you know too! I'll say it again: the real food celebrity is the farmer, our star in food—and the ABCSA is a celebration of their hard work in the culinary space too!

In this spirit and with the FarmOn! mission to stand up for our farming community and food choices, I approached ABC Kitchen with the idea to launch a city-friendly CSA program at the restaurant to help people meet their farmers and make a connection to food choices and sources in a way that is impactful and brings commerce to family farms in the Hudson Valley.

When I proposed the ABCSA instead of the traditional CSA program that typically offers many types of vegetables, Ryan Armstrong, general manager of ABC Restaurants, loved the idea and agreed to implement it immediately.

Having lived in Manhattan for many years, I understood all too well the mentality of eating in the Big Apple—small kitchens means little space, and dining out often is the norm. So together we launched the new concept of a city CSA and attracted over fifty-five members from the restaurant clientele, as well as actress Parker Posey, who signed on as a friend of the farmer with other notable celebrities and athletes. Members pick up their ABCSA curbside at ABC Kitchen bimonthly and learn why to skip the supermarket too. Same-day freshly harvested locally sourced food direct from the Hudson Valley farm to your table is divine.

Mindful of city life and culture, I wanted to give members options, so instead of a vegetable-only offer, I contacted local Hudson Valley farms and created two types of shares. The Farm Lover is a vegetarian option including organic salad greens; organic seasonal vegetables; fresh fruit; farmstead cheese from Hudson Valley creameries; a dozen eggs from free-running, pasture-raised chickens; and regional specialty food products. The Farmer's Choice includes each of these selections, as well as a mix of four pounds of grass-fed organic meat each delivery, which arrives frozen.

Ryan Armstrong said, "The ABCSA is the first truly curated 'citified' CSA created in New York City, complete with a gift each delivery, an onsite 'farmcierge,' great recipes from Chef Jean-Georges and NYC style

This page: FarmOn! partners with ABC Kitchen to bring the farm to your table with a twist on the traditional CSA offer called ABCSA, which is harvested the same day you receive your farm share. *Courtesy Cayla Zahoran.*

with a personalized messenger service. The feedback from our members is amazing each delivery, and the fun they are having with their share goes beyond just simple produce. It's really a part of a new connection to how food comes to the table, and we love being a conduit for more people in the city connecting to the great agriculture we have in this region."

Susan Brinson, an ABCSA share member, explained, "People buy into shares of crops in advance with local farms. For the farms, it's great. They know they're guaranteed to sell a certain amount of product and know how much to produce. For us, we get a box of goodies every two weeks. We don't know what will be in the box, and that's the part we really love."

One of my favorite farm-fresh recipes from Chef Jean-Georges Vongerichten also, I'm told, happens to be First Lady Michelle Obama's salad of choice when dining at ABC Kitchen. I'm sharing it here with you too!

Chef Jean-Georges's family-style kale salad at the FarmOn! annual Hootenanny! Benefit Dinner 2014. *Courtesy Cayla Zahoran.*

KALE SALAD WITH LEMON, SERRANOS AND MINT (SERVES 4)

By Chef Jean-Georges Vongerichten

ADAPTED FOR SERIOUSEATS.COM FROM *HOME COOKING WITH JEAN-GEORGES*

FOR THE DRESSING:
7 tablespoons red wine vinegar
2 tablespoons + 2 teaspoons lemon juice, freshly squeezed
½ piece of garlic, germ removed
½ piece serrano chili
5 tablespoons Dijon mustard
2 tablespoons kosher salt
Pinch black pepper, finely milled
1½ cups sunflower oil
5 tablespoons extra virgin olive oil

METHOD FOR THE DRESSING:
In a blender, combine red wine vinegar, lemon juice, garlic, serrano chili, Dijon mustard, salt and pepper together. Combine first set in vita prep and puree. Slowly add the oils together until emuslified.

FOR THE CROUTONS:
2 tablespoons extra virgin olive oil
sourdough-like bread, crust trimmed off, ⅛-inch cubes (slightly frozen, sliced on slicer)
kosher salt
black pepper, freshly milled

METHOD FOR THE CROUTONS:
Liberally coat the bottom of a sauté pan with oil and heat until smoking. Add bread and sauté until all sides of the croutons are golden. Place croutons on paper towels, and season gently with salt and pepper.

FOR THE PICKUP:
1 bunch Tuscan kale, washed and dried, ribbons
½ cup dressing
16 leaves mint, medium-sized, ribbons
3 serrano chili slices, rounds sliced extremely thin
24 pieces croutons
Lemon, to zest
Black pepper, freshly milled

METHOD FOR THE PICKUP:
In a bowl, mix the kale with dressing and mix well.
Evenly place the salad on a plate and top with mint, serrano
    slices and croutons.
Finish with lemon zest and freshly cracked black pepper.

Skip the supermarket produce aisle weekly and join a CSA. Community-Supported or Community-Shared Agriculture is also known as "subscription farming." It is sustainable, responsible and worth the effort. You invest with the farmer. The idea of "seed" money is invested as start-up capital you pay to the farm in advance of the growing season. This buys the farmer his seed and buys you a season-long subscription of food—just like you buy any subscription—but instead of receiving a magazine each week, you receive a "share" of fresh, locally grown or raised fruit and/or vegetables. Some farmers also offer CSA subscriptions for farm-fresh eggs, poultry, meats and dairy. CSA is a new name, but the process is reminiscent of an earlier time when people knew where their food came from, ate in harmony with the seasons and enjoyed delicious and healthy diets of pure, fresh foods. Sign up today with the farmers in your own community. To find a CSA in New York State, go to JustFood. org or Local Harvest.org nationally for participating farms. Or check out our guide to farms in the Hudson Valley in our resource guide. Once you find one easy to get to, just contact them and see if they have a CSA program that suits you. Chances are, they do!

"I'm lovin' it" should be the billboard for fresh food sourced directly from family farms. We shouldn't have to ask, how long has it sat there? How far did it travel in a truck? How did it ripen? Who grew it, and how on earth did it grow without seeds?

It makes me question the integrity of our "eat more plants" directive for better food and better health if the plants come from an industrial plant instead of the farm.

Local food is not only better for your health, but it also gives meaning and value to what we eat. Meaningful is all you really want when it comes to a good meal, so I'm getting you to dig deeper and explore all aspects of our food system. We all can't quit our day jobs and become farmers, but if you want, you can try your hand at farming without any permanent commitment.

At Kinderhook Farm in Ghent, New York, you can conjure up your own Green Acres dream, Arnold included!

Who hasn't dreamt of open farmland, rolling hills and fetching this morning's egg to scramble from the heritage hen just outside your door to—then you come to your senses and think that's a fantasy.

Reba the Great Pyrenees sheepdog watches over her grass-fed flock at Herondale Farm in Ancramdale, New York. *Courtesy Cayla Zahoran.*

But in the Hudson Valley, you can actually try your hand at farming, even for a weekend. Get the insiders' experience and all the bragging rights that go along with it on a farm-to-fork adventure. Book a farm stay in the beautiful barn on the property dedicated to the concept of sustainable farming, responsible land stewardship and the welfare of animals.

By pitching in for chores and tending to the animals, you gain an understanding of the relationship between the food you eat and how it is grown at Kinderhook Farm in Ghent. This hardworking farm team of seven people at the bucolic 1,200-acre farm in the Hudson Valley raises up to eight hundred animals some months that are all grass-fed, pasture-raised and certified as "animal welfare approved."

People say they don't want to connect the dots between the animal and their food when, in fact, it is exactly what you must do. As a result, you will change the way you eat and the future of farming for the health of our children and viable livelihoods in agriculture.

Kinderhook Farm is a collaboration between longtime friends Renee and Steve Clearman and Georgia and Lee Ranney. This year, they celebrate ten years of their vision and commitment to farming with their dedicated co-farmers.

The Ranneys had twenty years of experience raising grass-fed beef cattle in West Virginia before settling at Kinderhook Farm and are clearly skilled in their craft. The Clearmans made the venture possible as passionate partners.

Harry Lobdell, the farm manager who knows everything happening at all times on the farm (and who has amassed an impressive arrowhead collection from found treasures on the land), is the engine of the production.

Anna Hodson and Laura Cline are two savvy city women turned full-time farmers for professional apprentice opportunities, which has served them well by offering them exposure to farming methodology, training and sustaining profitability. Together, their work supports herds that are fed an exclusive diet of forages—grasses and legumes—which the species was designed to thrive on.

Lee's lifework is farming, but "Georgia is the queen and involved in all aspects of the farm," he exclaimed. "We also have very able co-farmers in our apprentices, so we don't have to do everything ourselves."

Proud proponents of the grass-finished food products, Kinderhook farmers also try to educate customers and visitors about chores, checklists and food choices to enable consumers to make their own decisions about grass-fed versus grain/corn-fed animals.

"I see us as part of a larger farming community that includes farmers who do things differently than us," Lee Ranney explained passionately.

Corn-fed beef contributes to a familiar texture and "mouth appeal" we are familiar with. While it soaks up salt like a sponge, when it isn't salted, corn-fed beef has the flavor of a sponge compared to robust, naturally flavorful grass-feed beef.

"Finishing" the animal by grass-fed grazing allows for how much fat cover and marbling is in your meat, which means tender, delicious meat. "Think of this inter-muscular marbling like pie crust where fat melts away and you get that flaky mouth feel and texture, making you feel satisfied," Anna Hodson told me. This is wholesome good food, period.

"It's a wonderful place to be and a wonderful place to live," Laura Cline tells me, gazing out from the farmhouse toward the land. A music and arts major who came to Kinderhook Farm by chance through web design, she continues, "You don't know at the time, but it feels like the thing we should all be doing. I love our animals very much. It's an active social space and a great summer destination spot with the converted barn for visitors and big open fields—you can take part and help or just relax on the hill and read a book."

Americans rarely eat lamb, and it is only those in the know who make it a diet staple. Once you eat grass-fed and -finished lamb from these farmers' hard work, it's unlikely you won't be a lamb lover. It is so extraordinarily tasty.

At the store on the farm, which is open seven days a week, you can buy all types of lamb cuts, hogget (considered a specialty—with only twenty of these animals on the farm at once—they are larger and fatter, so the flavor is more complex) and mutton (which needs a rebrand of the name!), which is simply a lamb raised two years or more.

Environmentally, Kinderhook Farm is using best practices on the land, implementing a closed system of grass and manure so there is no pollution and no runoff. The pastures are managed to optimize feeding by height and type of grasses eaten in rotation management to maintain the diversity of the grasses.

This makes all the meat it sells flavorful, and roaming allows the animals a better lifestyle. Despite the myth that grass-fed meat is tough, it's quite the opposite.

Bucolic landscapes and fresh air make a trip to the country revitalizing and enriching.
Courtesy Stephen Mack.

Being grilled on the open flame makes farm-fresh organic and pasture-raised chicken even more of a good thing. *Courtesy Cayla Zahoran.*

Grass-fed meat is lower in both overall fat and calories and has the advantage of providing more omega-3 fats. Animals at this farm are grass fed for various good reasons. The animals roam free in the fresh air, which is cleaner and better for the environment. They are healthier, too, as a result of their grass- and forage-fed diet, which is packed with nutrients and vitamins. The organic leaf of grass offers more nutrition than soy or corn. For more information on grass-fed versus grain- or corn-fed cows, check out www.eatwild.com.

When you think of chickens, you must differentiate between laying hens (for eggs) and meat chickens (for poultry). Knowing these birds will eat anything, they need a free-range environment to roam in the fresh air, never confined to cages. Pasture-raised, cage-free birds that feed on organic grasses eat a diverse diet in the sunshine and are less stressed and more flavorful.

Other practices punish chickens to meet consumer food demand, and we should make conscious choices so we clearly express what we want—and what we don't want—in our food. It makes what we eat healthy, so we look and feel better.

"Pigs come to farm this time of year and are heritage breeds like Berkshire. After spending time on five acres in open pasture, they feast on hickory nuts that fall from trees, which make them the local equivalent to Iberico ham," Anna Hodson tells me.

"Organic meat is expensive!" is often the cry of those not educated on the food system, so I asked Hodson her thoughts. Her reply was simple: "I feel that customers who come to our store are not rich. They choose to buy meat here because they believe in buying direct from our pasture-raised and grass-fed selections of meat and come to see the animals. We are very transparent. We even have vegetarians and vegans come and buy for friends because they trust us." She continued, "They know what they are getting, so they are willing to pay what is fair. It's about priorities—you have to want high quality."

Cline added, "We live like kings—we have everything! This should be a thing all people want."

It's so enriching to know that what you eat is also eating honestly and that the animals live so well you can honor this at your own table. Meet your meat—it does us all a big favor and ensures the animals are benefiting, too.

Hudson Valley landscapes beg you to take to the land. *Courtesy Stephen Mack.*

"BEEF & BEER" CHUCK ROAST (SERVES 6)
Recipe by: Paula Colarusso @thefoodishgirl

For years we have seen recipes that use processed short cuts to flavor cuts of beef like a chuck roast. Chuck is from the shoulder and has a good amount of connective tissue and fat that requires a good, long time in a slow oven for it to taste delectable. Skip the instant soup mix packet next time to season your next roast and instead try this: a savory blend of onions, garlic and beer to add flavor and tenderness to the meat. The resulting pan drippings are reminiscent of the best French onion soup you've ever tasted. The leftovers are amazing…as if.

INGREDIENTS
1 4–5 pound grass-fed local beef chuck roast, tied
Salt and pepper
Olive oil
4 cups sliced local onions
2 cloves garlic, chopped
1 pint Chatham Brewery Maple Amber Ale*
2 tablespoons Crown Maple syrup
1 teaspoon dried thyme

METHOD
Preheat oven to 300 degrees.
Pat your chuck roast dry and season with salt and pepper
In a heavy oven-proof cast-iron pot that has a lid, heat your olive oil over medium heat, and brown your roast on all sides.
Add the onions and cook for 5 minutes.
Stir in the garlic and then slowly add the beer. It will bubble up!
Add in the maple syrup and thyme. You may add a little water so that your liquid level is at least halfway up the roast.
Cover and put into the oven for 3–4 hours. It should be fork tender. Skim off the fat and serve the juices over the meat. Taste and adjust the salt and pepper before serving.
*Any type of brown or amber beer works the best in this dish. Remember that you want a sweeter, not bitter or hoppy, beer. Choose wisely for better flavor.

Chapter 6

SUCCESSION

Raise Your Replacement

Farming is as all-American as apple pie or baseball. Our agrarian identity is as red, white and blue as it gets. We believe in the American dream, and that, of course, involves opportunity and making money. But in the process of building our great nation, we somehow got away from our own roots. Suddenly, we are all realizing how far we have strayed from using seed and soil to honor, sustain and nourish our health.

It's hard to separate facts from fiction when all of our emotions are tied to the food we like to eat. It's family, tradition, place and politics all mashed together and spilling over the sides of our plates into and out of our hearts, souls, stomachs and minds. It's personal. You can't tell anyone what to eat or how and expect them to just pick it up like a T-shirt or a cool gym class and "just do it."

This much is clear: on all of our personal plates, we, as a nation, are starting to clear a space for farm-grown food. As highlighted previously, more and more Americans want wholesome food made by people in their own communities. Whether we're whipping up a grass-fed beef Bolognese, cranking out cabbage kimchi or just dipping freshly picked pink radishes in just-churned goat butter sprinkled with sea salt, Americans are choosing slower food.

But how much longer will we be able to all eat healthy, local food? And how can we sustain a growing, responsible food movement when we aren't sustaining our farmers?

There are 313 million people in our country, but right now, less than 1 percent of them identify themselves as farmers, according to the EPA. The scarcity of real farmers who make their living raising what we eat out of the ground is scary and seems downright un-American. Where would we be without our farmers? Who would feed us, and what would we eat? I can't imagine living on a diet based on food commercials, factories, fast food and frozen meals. Can you? It's depressing.

Despite the obstacles, there is a vibrant, grassroots locavore movement encouraging young, new farmers with fresh ideas with a passion for sustainability, responsibility and tech-savvy farm solutions, and it is our only hope in rebuilding our local food shed. We need our farmers to raise their replacements. We need a succession plan. As the founder of an educational farm through the FarmOn! Foundation, it's vital for me to support intergenerational farming in the Hudson Valley and educate and inspire new generations of would-be entrepreneurial farmers. I have also been working with local public school districts funding locally produced milk to serve during school lunch. The most important thing you can do as a consumer to ensure a successful succession plan is to increase your awareness of food issues and to go out and meet some of the farmers in your community, whether they've just planted their first seeds or have been making their living from the land for generations.

Some farming communities, like the ones in Columbia County, are creating opportunities for young would-be farmers and exposure for children who may not have ever considered the rural life but find they flourish in the pastures and fields. But some farm inculcation is happening on a more grassroots level, from the cradle on up.

There are a handful of farming family pioneers in the Hudson Valley, friends of mine who I've gotten to know in the fields and around our family tables, breaking homemade bread and swapping stories about growing up in farm life. All of these people have been working the land for generations and have raised their own children—their replacements—in such a way that they would grow up and want to supplant their parents behind the wheel of the tractor (instead of flee to the cities for a job that guarantees a salary).

Opposite: Sunset and the dog greet you after a long day of work and chores on the farm. *Courtesy Tara Boyles.*

Children learn to properly handle baby ducklings at FarmOn! *Courtesy Emmanuel Dziuk.*

This isn't medieval Europe. In contemporary America, it may not be the norm to pass on the manner in which we earn our money, our livelihoods, from generation to generation. But for these families of farmers, continuing in their parents' footsteps doesn't feel a constraint. It feels (and looks) a lot like freedom.

These farm families are in the trenches: they are showing us how to take back control of our land and our sustenance and revitalize a respect for farming with best practices, transparency and honesty in our food cultivation.

Big Food conglomerates spent $2 billion last year just to market junk food directly to their target audience: your kids. It is a really big business. "Food giants exist only to serve themselves, not the populace they are supposed to feed and protect," explained Chris Weaver, a local resident and MIT professor. "The locavore movement exists on many levels, with many agendas, but it is neither a political nor a liberal movement. It is a movement by the people and for the people that seeks to make evident much of what made this country great. By reexamining some of the things that we have forgotten, and recognizing the value of what these things brought to our grandparents and the generations before them, the time is now to restore that sort of benefit for our children and ourselves."

The more people who source food locally, the greater the demand. The greater the demand, the more people will be employed by farming. This creates jobs that are economically viable livelihoods. This also inspires farmers to "raise their replacement" for the next generation of the family farm instead of encouraging them to leave the farm for greener, more lucrative pastures in the big city.

In Columbia County, there is an amazing farm-to-school program called the HARVEST Club that is helping to show students every aspect of the farm life. HARVEST Club was formed by students in the Taconic Hills Middle School four years ago and has spread to encompass both elementary and high school now, with children involved from pre-kindergarten through twelfth grade.

Enrollment and interest continue to grow in the program led by Christie Hegarty, the district's agriculture teacher. Together with supportive parents and Director of Curriculum Sandra Gardner and her dedication to the HARVEST Club at Taconic Hills, the club even earned a national farm-to-school award in partnership with the Farm Bureau.

The school's programs include a full course study in agriculture. HARVEST Club endeavors include a "pizza" garden; a hard-top greenhouse; raising pheasant, chicken and lamb; and berry and pumpkin planting. Students learn the business of food, fundraising, caretaking, responsibility and marketing all necessary for viable livelihoods in agriculture and succession on family farms.

HARVEST Club kids are talking about tools for building sustainable local economies from farming initiatives, and Ben Gardner, vice-president of the HARVEST Club, told me, "We try to be good role models, much more conscious of what we eat, partly influenced by the club and also by hearing about the problems in food—people in our school are spreading the word—peers, coaches and teachers. We try to drink less HFCS [high-fructose corn syrup], but if the milk doesn't taste good, kids won't drink it."

Ian Perry, a founding member of the HARVEST Club, continued, "I like Ronnybrook milk. I wish we could change our milk at school. The one we have, I looked it up, it has additives and I stopped drinking that. I want to know if Ronnybrook or Hudson Valley Fresh can sell milk to my school? Maybe with milk dispensers of local milk instead of cartons? I trust our lunch lady Mrs. Strompf is feeding us the best food she can."

Still smiling after a long day on the tractor, Bob Fix is the patriarch of the multigenerational Fix Brothers Fruit Farm in Hudson, New York. *Courtesy Cayla Zahoran.*

At home, Ian raises egg-laying chickens as pets and invented a hydroponic growing and composting system. "Just buy one, add water, everyone can grow their own food," Perry said enthusiastically.

Karyn Johnson recently moved to the district. She is eighteen years old from Philmont and excited to be a part of the HARVEST Club, as her dad had been a farmer. She said, "It is pretty cool to be able to learn about farm animals. I can be a part-time farmer running a vegetable stand [along] with my chosen career plan to become a police officer."

Johnson's grandmother gets her vegetables out of her garden without the use of any pesticides, and the family shares meals together. So naturally, Karyn enthusiastically joined the HARVEST Club at Taconic Hills when she moved to the area.

Her motto? "Hey dirt, let's go!" We can all learn a lesson from this idea!

At a time when most kids are uninvolved or distracted with passive entertainment pursuits, these ideas of growing the food you eat, caring for animals, opting out of processed foods and making pesticide-free food choices are refreshing. It's a view to the future of farming from hardworking kids who think like businesspeople.

Ian told me, "If I can make money in ag, I want to stay here. I love it here and want to be an engineer in ag and develop solar-powered tractors that gain energy plowing fields in the sun."

The McEnroe Organic Farm in the Hudson River Valley in Millerton is one of my favorite family farms, and every year, without fail, I buy my Thanksgiving turkey there.

As legacy goes, Ray McEnroe had three sons, and the McEnroe original family farm dates a century back, having been located in Amenia. The patriarch made a living in real estate and harvested iron ore at the time on family land.

Ray McEnroe II was a progressive dairy farmer and invested in a herd in 1951 where the Millerton farmland is today. His son, Ray McEnroe III, grew up farming and was friends with Ronny Osofsky of Ronnybrook Farm—the two of them have colorful stories about "million-dollar cows" sold at the Madison Square Garden Dairy Sale—back when folks still invested in agriculture as the real future of farming. These families are the real deal in raising your replacement, inspiring their kids at an early age

Organic pasture-raised white broadbreast turkey at McEnroe Organic Farm provide you with the best meal on the planet you will give thanks for this Thanksgiving. *Courtesy Erich McEnroe.*

about the business of food and the importance of farming to sustain the family business that sells to the Big Apple as a major source of revenue. It keeps their children interested and earning.

In the late 1980s, Douglas Durst bought sixty-eight acres of land from Ray McEnroe III and launched a soil business with an elevated vision of compost as an alternative to landfill in years to come. This intrigued McEnroe III, who partnered with Durst, and it evolved from there.

Today, they continue to grow the business and produce high-quality organic soil, making dirt look good—and feeding you better from it, too.

Since 1621 when the Pilgrims arrived in Plymouth, Massachusetts, tradition dictates that turkey is the centerpiece of Thanksgiving, the celebratory day of harvest, abundance, freedom and gratitude. And a turkey from McEnroe family farm will make your family and friends grateful, too—some years, I even go and meet their turkeys and watch them frolic in the fields before Thanksgiving. I have always thought meeting your meat makes you appreciate it more.

Baby turkeys, or "poults," begin arriving to McEnroe farm in mid-June through the end of July (that's how you get pound variation, the average being sixteen to twenty pounds but ranging from ten to thirty pounds for a few feasts!) when they are one year old from a family-owned hatchery in Pennsylvania.

They start summer life in kiddie pools for two weeks, which are elevated on tables so they are easier to maintain, and the edges of the pool stay round, not sharp, so the poults don't huddle and suffocate. They live under heat lamps at ninety to ninety-five degrees Fahrenheit to keep them warm and happy. They are then sorted into groups and live in the barn for two to seven weeks, until they are weaned off the heat lamps as their feather coats develop. Then they roam the pasture outside on the farm and grow strong for four months in the fields in moveable pens, feeding on organic alfalfa, which provides protein and roughage. They rotate every ten days on different fields, grazing on forage and eating less grain to grow at a better rate and naturally

fertilize the fields for next year's grain crops (typically corn, a nitrogen-dependent crop). This makes the turkey a dual-purpose animal for fertilization and meat sales.

When it finally gets cold at night, they move inside and gain all of their juicy weight in the last month. At McEnroe's, they sleep comfortably in cattle trailers from dusk til dawn to avoid predators and stay safe from any heavy rain.

Reservations for the birds start on Labor Day,

Greenhouses growing organic tomatoes at McEnroe Organic Farm in Millerton, New York.
Courtesy Cayla Zahoran.

and with a limited number available, you have to be sure to get your order in early!

Erich McEnroe is one of five brothers (Wade, a talented chef, prepares food in the café year round, including all the fixins for your Thanksgiving and his own family dinner; Kyle lives in New York City; Ryan is in D.C.; and Sean is in Colorado). He is the next generation to heed the calling to the land as Ray McEnroe III raises his replacement. Erich works alongside his dad on this 1,200-acre organic farm raising livestock of pig, cattle and lamb that are free range and grass fed on 400 acres, as well as chicken and turkeys that are, of course, seasonal.

Since I get my (always delicious) turkey from the family every year, I decided to share my secret with you, as well as a side dish to savor from a friend of mine @thefoodishgirl.

McENROE THANKSGIVING HARVEST GRATIN (SERVES 6)

Recipe by: Paula Colarusso @thefoodishgirl

INGREDIENTS
Light olive oil for frying
1 medium white onion finely chopped (about ¾ of a cup)
1 large garlic clove smashed and minced fine
3 cups potato (preferably a yellow-fleshed Yukon) peeled and cubed
3 cups pumpkin (a small pie pumpkin is perfect or substitute butternut squash) peeled and cubed
½ teaspoon summer savory (can substitute thyme)
1 cup Beechers Jack cheese grated
1 cup grated parmesan split into two portions
1 pint Hudson Valley Fresh heavy cream
½ cup chopped pecans
½ cup chopped dried cranberries

METHOD
Over medium heat, fry the onion and the garlic in the oil until translucent.

Remove from the pan and set aside.

Into your hot fry pan, add the potato and pumpkin and fry until browned on a few sides. This will take a few minutes; be careful not to burn. Once browned, remove from heat and add the onion/garlic mixture. Season with salt and fresh black pepper and ½ teaspoon of summer savory and stir in the jack cheese and ½ of the parmesan. Toss to coat, then tumble into a greased ovenproof casserole dish. (At this point, this dish can be covered and refrigerated for up to one day before baking.)

Pour the heavy cream over the mixture, top with the rest of the parmesan (use a little more for good measure), the pecans and the cranberries. Bake uncovered at 350 degrees until browned and bubbly.

ADD-IN OPTIONS:

Add crushed hot red pepper and a shake of smoked paprika instead of summer savory and top with garlic-buttered bread crumbs instead of the pecans and cranberries for a more hearty take.

When the holiday feasting is over every year, as much as I love the pickles and preserves, winter squash, snow-covered Hudson Valley hills and dashing through the snow, I start getting little cravings for fresh, sweet, local fruit. As soon as the first blades of grass can be seen peeping through the snowdrifts, I'm piling myself and my bulldog in my truck and hitting Kleins Kill for my fruit fix. (And yes, it is so worth the wait between fruit harvests—it actually makes you eat in season more!)

Kleins Kill Fruit Farm feels more like an office on Wall Street, bustling with movement and sales, than a fourth-generation regional family farm. It's all about bringing fruit to the public to meet the needs of the fresh and value-added marketplace. So the window is short, and the pace is fast.

Meet the Hudson Valley "Fruit Mafia"—the legacy of Antonio Bartolotta—entrepreneurial, honest, warm, generous people who work hard to maintain family values and compete in the big business of fruit

production and sales in New York State with their business: Kleins Kill Fruit Farms Corporation of A. Bartolotta & Sons in Germantown.

Six brothers (Benny, Tony, Russell, Al, Phillip and Robert, who range in age from eighty-six to seventy-four years old) and a brother-in-law (Armand Conte) made this business a wholesale success story buying and selling apples that are an economic driving force in New York State today.

Their family business is one of the largest in the region, growing apples and feeding folks at supermarkets and restaurants with ready-packs pre-sliced for kids.

The family that farms together stays together. When I visited Russell Bartolotta Jr. at his farm, there were one hundred employees busy harvesting—a whole lot of togetherness! But I left not only inspired by their hard work and camaraderie but also in the know about fruit economics and what it takes to make that peach perfect.

That's the thing about perfect fruit: it's rare, Russ Jr. told me. "It's all about timing and the variables that can put you out of business in a flash," he said. That's why on this family farm, "it's a fascinating business.

Never a dull moment. No matter what, there is always one 'girl' you will have to worry about—even when you grow, ship and market yourself—and that's the wrath of Mother Nature."

Russ Jr. explained further, "The Northeast apple business has been hell for the last thirteen years. Between hail and frost, if you cannot afford to implement different technologies as an

Right: The perfect peach from the Hudson Valley—tree ripened and local. *Courtesy Tessa Edick.*

Opposite: Organic tomatoes make for great juice, salsa and tomato sauce at McEnroe Farm Market. *Courtesy Cayla Zahoran.*

Farmer Bob Fix on his six-generation family farm with fruit trees in bloom. *Courtesy Linda Fix.*

investment, you won't make it in farming as a mid-size farmer in fruit. But if you can, and Mother Nature cooperates, it's a really great way of life."

Luckily, we live in New York State, which means we are the second biggest producers of fruit in the country, only second to Washington State. And as nicknames go, we live up to our Big Apple reputation. At Kleins Kill Fruit Farms, the apples are so perfect you won't want to eat just one a day. The sweet juice that drips down your face after one bite of a peach brings a joy that far surpasses the typical supermarket experience with fruit.

"It's a lesson in economics, this business—basic supply and demand principles that make opening day of the harvest season my favorite day of the year," Russ Jr. told me with smiling eyes. Thanks to the impressive business acumen that has kept this family in the fruit business for generations, with big plans for growth in the future, Russ Jr. and the rest of his family should have many more opening harvest days to celebrate.

This family doesn't only have longevity but legacy, too, with an Italian heritage (and spirit!) descending from Sicily. In 1904, Antonio Bartolotta came to Ellis Island and worked on the railroads his first few years in New York City until he arrived in Columbia County in 1912.

In the fall of 1912, he was working for a farmer named George Van Dyke on the farm of Russell Cooper, who could trace his ancestry to the blue-blooded Livingston family, the most powerful family in the area during the colonial era and beyond.

"Antonio fell in love with the farmer's daughter," said Robert Egan, one of Antonio's grandsons. In 1913, Antonio married Jessie Cooper, and they had thirteen children (who produced forty-three grandchildren and fifty-five great-grandchildren) who mostly live in the area today around Linlithgo, a hamlet of Germantown.

In the spring of 1921, Antonio and Jessie Bartolotta bought their first farm and called it the Home Farm where the homestead was, which was 225 acres. They renamed the farm Kleins Kill Fruit Farm because of the Kleins Kill creek, a tributary of the Hudson River. The word "kill" means creek in Dutch—the original settlers to agriculture in Columbia County.

"This isn't something you do unless you have passion and love it— better to grow up in it than learn on your own," Russ Jr. told me. Four generations later, they have a formula to sustain profitability that is at work and working.

Buying and selling apples as a vertically integrated business, this family farm sells coast to coast with the Bartey Brand growing annually as big retailers look to connect consumers to the family farm for better, nutrient-dense food choices.

Russ Jr. graduated from the University of Massachusetts–Stockbridge in agriculture and ran the family farmstand in Rhinebeck until 1999. In 2000, he took over operations on the farm and encouraged his son Zachary as a teenager to foster a love for fruit farming, too.

Zach, a student at University of Massachusetts at Amherst, is following in his father's footsteps. Russ's daughter Allie is also studying agriculture from the business point of view, and their cousin Adam works at the farm helping coordinate food safety measures for Good Agricultural Practices (GAP) certification.

Each year, the family plants new trees in April, grows crops from April through July and harvests peaches, nectarines and plums until mid-September. Early apples and pears harvest mid-August

Early spring apple on the tree. *Courtesy Cayla Zahoran.*

through October, and once the plants go dormant in November through the winter, the family begins trimming and prepping for the following year.

The original farm harvested many currants and grapes, as well as gooseberries, sour cherries, peaches and apples in varieties that have long gone to rest, including "Black Twig, Baldwin, Northern Spy and Rhode Island green," said Russ Sr., describing the old-time apples.

The new hopeful is the early apple called Zestar, a great eating apple (tastes like Honey Crisp mixed with Macoun), but it hasn't yet replaced the favorites of Russ Jr. (Empire), Russ Sr. (Macoun), Al (Cortland) and Robert (Gala).

With new technologies and retailers no longer refrigerating apples, the business moves fast. You have to stay informed, planting higher-density, higher-yield trees to maximize output and be ahead of the curve, which is why at Klein's Kill, fifteen thousand to twenty thousand new trees are planted each season, replacing old trees that are then used for firewood.

Higher-density, new trees yield more fruit, and costs are lower. "It's a world market now, which means I'm out hunting for new business every day. What happens in Europe or South America affects our nation, and the apple business is a great economics course," Russ Jr. tells me passionately while perusing the *World Apple Report* and noting, "China grows four times the amount of apples, but 78 percent are Fuji—but since they don't have the right root stock or storage they cannot diversify the crop, so we as an industry must be on the cutting edge."

An apple a day really does keep the doctor away. You can live a great long life, too—just ask: who makes my food? And don't be cheap! If you are cheap with your food, you are cheating your health. So be picky and get pickin' family-farmed apples this fall. Your own family will thank you—and the Bartolottas will too.

The next time you want to celebrate your American roots and a local farmer, think of the Hudson Valley fruit farmers or, better yet, visit one. Fix Brothers Fruit Farm offers Pick Your Own (PYO) annually September and October at its Hudson-based farm.

HOMEMADE FARM FRESH APPLE PIE (SERVES 6)
Recipe by: Paula Colarusso @thefoodishgirl

Almost one hundred years ago, the art of pie baking was lauded and exalted, though there was a dark time that held that fried pies (apples stuffed into a folded-over pie crust and fried in fat in a hot iron skillet) and even the fat in the pie pastry could render a pie "indigestible."—Horrors!

Luckily, that didn't deter our ancestors from combining fresh or home-canned fruit with delicate pastry to the delight of generations to follow. One thing missing from cookbooks of today is the word *taste*... it is there as a noun. It should be used as a VERB much more often, especially when using fresh local ingredients.

BE CHOOSY!

Choose your apples carefully. It doesn't matter if there are scars, pits or blossom end strangeness. You can always use a paring knife and cut out anything unsightly. You will want a tart apple for your pie. In the days of yore, a homemaker could choose from Rhode Island Greenings Gravensteins or Yellow Porters for their pie-making pleasure. Now your selection could run Jonagold, Pippin, McIntosh, Granny Smith and Fuji. All hold up nicely to the heat of the oven. Mix and match. Each apple, like a vintage of the finest wine, has a unique flavor and texture.

INGREDIENTS:

4–6 medium-sized apples cored, peeled and cut into eight pieces (or thinner if you feel choppy that day). The goal is to have a heaping selection of sliced apples in your pie tin. You might need to add more; use your empty pie tin as a guide.

½ to 1 cup sugar

skip this next step if you are a purist and believe that a true apple pie is unadulterated with spices of any kind

½ teaspoon of mace (super old school)

½ teaspoon of cinnamon with a scrape or three of fresh nutmeg
Choose one or none.
¼ stick of local butter cut into pats

METHOD:
Preheat oven to 450 degrees and move baking rack to bottom of oven.
Start by TASTING your cored and peeled apples! Place them all into a large mixing bowl.
Add the sugar, mix and taste again. You can always add more sugar; you can't remove it. TASTE!
Add spices as you like.
Place the sugary apple mixture into your pastry-lined pie tin.
Dot top with pats of butter and cover with your top pastry, seal and pinch the edges. Cut out a hole in the center of the crust, the size of a nickel, for steam release. Brush the top crust with a tablespoon of fresh milk and dust with sugar.

Place the pie in a 450-degree oven for 10 minutes on the bottom baking rack. After 10 minutes, adjust your temperature to 350 degrees. Remove the pie to a hot pad, carefully move the baking rack to the middle of the oven and put the pie back in and continue to bake at 350 for 40–50 minutes. The pie is done when you see and hear bubbles, have a lovely browned crust and when a sharp knife dipped into the center of the pie will reveal tender apples.

PIE CRUST
The Foodish Girl's mother and her seven sisters each had their own variation of pastry recipes. Each lady carried with her a unique method and tradition either found in a church cookbook or a cookbook that came free with a tub of shortening. Either way, pastry dough is easy and quick to prepare. It does take practice. Nothing is better than one that you roll out yourself.

INGREDIENTS

2 cups sifted flour

½ teaspoon of salt

¼ cup butter

¼ cup vegetable shortening

1 cup of water with ice cubes—must be COLD with a
measuring spoon handy

METHOD

Sift your dry ingredients together and put into a mixing
bowl.

Add butter and shortening and, using a pastry cutter or
two knives, cut the shortening into the flour until you
have a slightly moist mixture that looks like beads of wet
sand.

Carefully add 1 tablespoon at a time of ICE COLD
WATER, using a rubber spatula to scrape the bowl and
gently turn the mixture onto itself. Add the water JUST
UNTIL YOU HAVE AN ALMOST ball of dough. It
should still be slightly moist, not sticky, and you should
be able to squeeze it into a ball. If sticky, you added too
much water...add a little flour and gently mix again.
Stop.

Remove the ball of dough and cover with plastic. It
needs to rest for 10 minutes.

Call your friends and tell them you are making pie.

After you get off the phone, cut the dough ball in half
and place on a floured surface (your kitchen counter
will do) and roll out with a floured rolling pin. Roll
away from you in firm, short strokes until you have a
thin disc of dough. Remove from the counter (use a
knife to help it up) and place into your clean pie tin.
Follow the rolling directions for your second piece of
dough for the top crust.

You will have extra dough from the sides where you trimmed
and sealed your pretty pie. Roll that out, brush with milk, add
cinnamon and sugar, place on a baking sheet and bake in the
oven for a few minutes. It's a delicious snack!

The Oomses are another example of the too-rare breed of Hudson Valley family farmers, and they work as hard at ensuring the future of family farming as they do at maintaining their own.

Like the Bartolottas, the Oomses believe the family that works together stays together, and at this second-generation dairy farm, you can certainly feel the love among nine employees who work seven days a week to feed us and their cows as wholesomely as they can on their 1,500 acres of farmland.

"Comfortable cows make better milk, and if you don't take care of them, they won't take care of you," Eric Ooms tells me on the way into the parlor for milking time.

It's 3:40 p.m. and the second milking at A. Ooms & Sons. I met with Eric and his brothers Tim and Ron at their farm in Valatie. Time isn't something you take for granted at the farm. With a daily wake-up call at 3:40 a.m., 100 heifers and calves to tend to and 350 cows to milk twice a day taking up to three to four hours a go (ten-minute milking time per cow), organization and sanitation of the ladies is crucial.

Every morning, the ladies are ushered into their milking stalls from their freshly groomed and sawdusted beds, and each day's milking yields a total of nine gallons of raw milk per cow. These ladies live a noble life—comfortable, clean and happy as a herd. With transportation of fresh milk every other day from the farm, Eric explains, "Highly perishable fresh milk has to move quickly." That means the milk is loaded with nutrition from their family to yours.

With tight control on costs and premium-quality milk that the Oomses sell wholesale to Agrimark, McAdams, Cabot and Beechers, you wonder how much can they earn and why does the entire family stay in the dairy business. Most importantly, is it sustainable?

The answer is simple: even though the job entails long hours and hard work, they do it for you—for your health, the health of your children and the future of this country—which is also why Eric Ooms is vice-president of the New York Farm Bureau.

He intimately understands the legislation needed to sustain the business of farming in our own region and nationally dealing with tax assessments, minimum wages and responsible farming practices. "Too many farmers are negative about farming and that it doesn't make money, but the truth is who makes that much more money in a cubicle in corporate America anyway? I love my job. People have to be happy."

Luckily for the hardworking Ooms brothers, their entrepreneurial native Dutch parents, Adrian and Dinie Ooms, emigrated from the

The Hudson Valley landscape is magical at dusk and dawn. *Courtesy Stephen Mack.*

Netherlands in 1950. They met at a square dance in Altamont and bought a farm in Old Chatham in 1952. They raised their five children on the dairy farm and in the farming business they started.

Living the farm life gave these children a head start, and the boys grew up learning the business. "If you can communicate, you can do anything," Eric told me. "My father made a great living on fifty cows with a pasture-raised grass-fed program for five months a year, but feeding five hundred grass-fed cows in the Northeast is challenging. In order to keep them clean, healthy and fed in the off season, we raise as many cows as we can handle, and it is working well for us now."

In 1982, Adrian Ooms bought a second farm in Valatie where production runs today with the idea that with four sons, it would be easier to milk cows in two places. Uncle Tony also has a nearby farm he started in 1978 called Oomsdale Farm. Randy Ooms left for northern New York and bought a dairy farm near the Canadian border in Malone, where he now milks a herd of seventy cows. One hundred and eighty acres of the original farmland in Old Chatham was donated to land conservation, and today Eric; his wife, Catherine Joy; and their three children live in the same house on the property where he and his brothers were born and raised. His sister Cindy lives next door.

Food is all about relationships, and these brothers work amazingly well together given their 24/7 work schedule tending to the rigorous daily routine at the farm. They rotate work schedules on Sundays for family time since they work one hundred hours a week each. They grow 80 percent of their own feed for the cows, which includes silage (first-cut grass from last year that is fermented) and grass from the third cutting last year, as well as corn silage (from corn they grow and crush the stalks).

Tim Ooms is the engineer and makes sure all of the Concentrated Animal Feeding Operations (CAFO) and Comprehensive Nutrient Management Plans are kept up to date. It's a lot of work, but the state requires it, and this strategy for keeping the farm clean and profitable is working.

Today, with 450 cows on the property that produce hundreds of gallons of fresh milk, it's a kid's paradise and the way life should be. There are a dozen show calves, and one in particular has quite a personality. Alpo loves people—but at 1,400 pounds, you need to be careful when playing!

At just fourteen years of age, Emily Ooms (Tim's daughter) is already active in agriculture. She's a Dairy Ambassador for Columbia County, a member of the Eastern New York Holstein Club and is tractor-certified from a course at Wil-Rock. She participates in Dairy Visions to visit farms, helps Dad milk some Sundays and takes care of the calves all of August to ready and show her "girl" at the Chatham Fair. In 2012, she took home first prize as

A Hudson Valley ear of corn not yet harvested from its stalk.
Courtesy Cayla Zahoran.

Bethany Meyers of A. Ooms Dairy with her dairy cows. *Courtesy Cindy Meyers.*

"Grand Champion of Holsteins" at the 4H show—with a grand ribbon to prove it! She doesn't know what she wants to do for work, but if she ends up back at the farm in some capacity, she will surely be an asset to the team.

Eager to visit farms and foster the entrepreneurial spirit to pair with the family business, Emily and her cousin Katelyn Ooms (Ron's daughter) participated in Camp FarmOn!, a fully funded "Homegrown Business Challenge" summer enrichment camp with Cornell Cooperative Extension, Questar III and 4H, and presented a profitable idea for a local farm to a "shark tank," showing off their business acumen and confidence in having an influential role at the farm one day.

Emily told me, "I never drink store-bought milk—I always drank raw milk from the farm. I tried store bought once, but it didn't taste good—it tasted watery." Clearly, nutrition tastes good and is beneficial; Emily is healthy, smart, gorgeous and affable. Eric told me, "The reason people started pasteurizing milk was to make sure it was safe." That doesn't mean it is necessary or more nutritious.

Support the Ooms family farm and their hard work by buying the eighteen-month aged cheddar available at Beecher's Handmade Cheese in the flatiron district in Manhattan or online at BeechersHandmadeCheese.com.

Their formula for aged cheddar is a mix from Holstein cows the Oomses raise and milk from Jersey cows from another farm down the road. Kurt Beecher Dammeier, founder of Beecher's Handmade Cheese, believes that truly great food starts with local ingredients and adds, "As a cheese maker, we believe that the quality of the milk has a direct effect on the quality of the cheese, which is why we work only with locally sourced milk from dairy farmers who take great care of their animals, consistently providing pure, great-tasting milk free of added hormones."

MAC AND CHEESE RECIPE (SERVES 2)

Recipe by: Paula Colarusso @thefoodishgirl—an adaptation from the 1950 Betty Crocker Picture Cookbook
This is one of the easiest recipes to assemble and bake. You can easily double this recipe!

INGREDIENTS:
8 ounces (dry) macaroni, cooked al dente
1 cup Beecher's Flagship cheese grated
1 cup Beecher's Just Jack cheese grated
½ stick butter
2 cups fresh local quality whole milk
¼ teaspoon of cayenne
¼ teaspoon of dried mustard

METHOD
Preheat your oven to 350 degrees.
With a knob of butter, grease a 9 x 13 pan.
Layer the cooked macaroni and cheese into the pan with pats of butter on the cheese. (Macaroni, cheese, butter, repeat! Save a few pats for the top of the dish.)
Stir the cayenne and the mustard into your milk and pour over the macaroni.

Bake until the edges are golden brown and you see bubbles—remember, the pasta is already cooked so don't overbake.

Add-In Options:
Add farm fresh vegetables, sausage, lobster or crab.

Chapter 7

COMMON SENSE

Honest Food and Responsible Eating

Who feeds you? Do you know that hundreds of small-scale, multi-generational family farmers are forced to leave their land and their legacy every week because they earn less than what it costs to make the food? Do you know that the nutrition needed to supply physical growth and good mental health is only consumed by 21 percent of high school students and directly leads to poor academic performance?

What is America eating? What is so super about the supermarket? When did we swap nutrition for convenience? What is healthy? What is eating right? Is there a perfect food formula? And why are we feeding the kids, the elderly and the sick unhealthy food?

Ask your mother, a nutritionist, Oprah, an architect, an athlete, a farmer or yourself, and chances are, you'll get many insightful answers.

Truth is, it's subjective. No one wants to be told what or how to eat, and answers to these questions are indeed debatable depending on your mood, your location, your health, time of day and with whom you are eating. In fact, you can't tempt yourself to like something you naturally disdain, no matter how healthy, how much your best friend loves it or—sorry, Julia Child—even with a bucket of butter. What we all *can* choose *almost* all of the time is food from the farm, something wholesome and uncontaminated and from sources that have put sustainable practices in production.

Making food responsibly and practicing land stewardship through quality production and transparency as to how it's grown, what it's fed

Hand-crafted dried sausage and other fine charcuterie made locally by Jacuterie in the Hudson Valley. *Courtesy Cayla Zahoran.*

and the quality of the air, soil, water and seed that the farmer uses to make it all possible are vital to eating right and living healthy.

I don't believe in diets. I believe in honestly made food. I believe in succession on the family farm for preservation of our local food system. I believe in cultivating biodiversity and terroir for a sustainable future, quality food production and decline in illness. I believe meeting a farmer who feeds you makes all the difference in the world for your well-lived, balanced and fulfilling life.

I educate myself about where my food comes from and make it a habit to ask questions that lead me to a food way that is sustainable. It is my food formula and a thought process that becomes a lifestyle, naturally and organically connecting people and leading us all to health and wellness globally.

Eating sustainably means the opposite of depleting natural resources, which is our food production habit to date in this country. When you source food directly from local farms you make a difference beyond your plate: you maximize your nutritional intake *and* small farmers' profits. To ensure succession and viable livelihoods at these family farms, it will take time, especially in a challenging environment (economically and politically)

Dr. Oz and Tessa Edick stand up for local food and family farms in the Hudson Valley. *Courtesy Emmanuel Dziuk.*

with government bureaucracy and well-funded lobbyists invested in the business of food that makes you sick. It is far more profitable to fix you with prescriptions than prevent illness with nutrient-dense food. And it all comes down to choices: which food do you choose to spend your food dollars on every time you eat, and where does it come from?

Rarely will a farmer get rich from making your food, but somehow in the big business of feeding, corporate farms are getting rich at the expense of our health and promoting cheap food by way of clever Madison Avenue Mad Men's advertisements packaged with jingles, tag lines and promises for a happy life. Shortcuts and slick slogans have tricked us all, and fast food has rendered us fat. Wake up, America—I'm here to convince you all there is another way to eat! In the long run, these empty calories that sold happiness and the promise of superior taste only jeopardize our longevity and increase the rate of cancer. This toxicity leads to economic depression and further disengages people from the farm and its wholesome food. Even in marketplaces where rural agricultural communities are located within one hundred miles of a major metropolis—like the Hudson Valley, nestled between New York City and Boston, Massachusetts—people don't know what or how to eat local by locating a farm or its importance. It's confusing, it's complicated and it's time for change.

With access to apps, recipes and the Internet in our pockets, how many times have we all stood in the supermarket assiduously reading labels like the good consumers we are, wondering: What is all-natural? What is homemade? What are trans-fats? What is USDA organic? And isn't pasta sauce already gluten-free?

Skip it, I say. They are indeed tricking us with labels, trends and expensive packaging that promises a long shelf life and better value instead of value for the farmer who feeds you. Instead of becoming a food label decoder, follow one rule: meet your farmer who will show you how to eat local when you can and won't dupe you into eating food that is making you sick.

If you look at hard statistics on farm profitability and who's making the money, rarely is it the small-scale farmer. There is very little reliable data available about local farmers, but even the picture for the big guys

Opposite: Undo the tricks of the trade and commit to eating farm fresh food in season. That's sustainable. *Courtesy Cayla Zahoran.*

is getting less and less rosy every day. According to the USDA, net farm income is forecast to be $95.8 billion in 2014, down 26.6 percent from 2013's forecast and the lowest since 2010. Lower crop cash receipts and reduced governmental assistance are the culprits, the USDA reports.

When you buy your food locally direct from family farms (or supplement and grow it!), you gain control of your health and switch up the game. Spend the food dollars you earn at the family farm and trust that farmer to feed you and tell you the truth about best practices. Start one conversation that binds communities and triggers commerce, leading to economic development in your neighborhood, community garden or even your own backyard with the return of the victory garden.

It's a simple food habit we can all cultivate a taste for—a formula, if you will—happening all over the world and for centuries based in agrarian roots that stimulates growth. In the Hudson Valley, all of this awareness is starting to have a serious impact. According to the Glynwood Center in Cold Spring, a working farm and sustainability think tank, the average direct sales from farm to consumer are 52 percent higher in the Hudson Valley than in the rest of the state. That means more people here are aware of the titanic wave of impact even the most minor purchase can make on their community. They are shopping for their produce and dairy at farm stands, buying meat from farmers' markets and becoming shareholders in CSAs. More people here are inspired, and more people are returning to the Hudson Valley to farm than in other areas of the state and the country.

But we're not in the clear: between 2002 and 2007, there was a 3 percent drop in the number of farms in the Hudson Valley, a 21 percent loss in the number of farms with five hundred acres or more and a 21 percent increase in the costs of production, Glynwood reports. And more than half of the farmers they interviewed in 2007 (the latest data available) didn't even work full time as farmers. They either worked two jobs or just farmed into retirement.

Rebuilding America's economy with family farm–centered food systems is at the core of FarmAid's standalone concert mission. The president of FarmAid, Willie Nelson, explained it simply as we turn to local food for sustenance: "In 1985, we started out to save the family farmer. Now it looks like the family farmer is going to save us. As our nation continues to endure an historic economic downturn, America's family farmers offer us much hope."

So while buying local may present logistical challenges (we all want strawberries in February!) and may require a few extra steps here and

Pigasso Farms Market selling this morning's eggs, poultry, pork, lamb and beef in Copake, New York. *Courtesy Cayla Zahoran.*

there, you can feel good about the extra effort, knowing that you are supporting families who are genuinely struggling to put food on their tables so they can put food on *your* table—and you can leave the fat cats in the big business of food to their chemical-laden boxed treats. Their pockets are deep enough to spare the financial hit.

An ancient Ayurvedic proverb reads, "When diet is wrong, medicine is of no use. When diet is right, medicine is of no need." And why would we rely on medicine to fix ailments that can be prevented with food choices from reliable sources?

People say they can't afford to eat locally, but I think they need to analyze it a little differently. What does it cost to be sick? To miss time from work? To pay for a doctor's visit? What do you pay for supplements and prescriptions? How much do you spend in remedies for stomach upset, anxiety, allergies, weight loss or vitamins and supplements because your food simply lacks nutrition?

People need to take one simple step: Go to the farm and come to an understanding of what real food actually is, how it's produced and what it does for your body. Would you put cheap fuel into a Ferrari and expect it to offer a top performance? Then why wouldn't you invest in the best "fuel" you can for your body to optimize and function at the highest level possible for good health?

More than one-third of adults, or about 72 million people, in America are obese, according to the Center for Disease Control. And in the Hudson Valley, land of hills and valleys? It's even worse: 57 percent of adults in the Hudson Valley are obese or overweight, according to an analysis by Senator Kirsten Gillibrand. Obese people incur about $1,429 more in medical costs per year than people of normal weight annually, and the annual expenditures for obesity-related diseases in America are estimated to be as much as $147 billion, the CDC reports. That's right: *billion.* Suddenly, that extra ten miles to the farm stand instead of ten minutes at the drive-through doesn't seem as inconvenient or costly to look and feel great. There's a remedy in food that treats almost everything that ails you.

When you practice prevention with nutrient-dense food, you alleviate the hassle and worry of illness requiring doctors and medicine to treat the effects of bad food. We have to start to reexamine our food system and our own personal consumption, what we eat and where it came from. Even the honeybees have started to flee, they're so sick of our antics, antibiotics and overuse of chemicals.

While I don't diet or overeat empty calories from processed food, I do follow the seventy/thirty food rule. It leaves plenty of room for vices—70 percent of the time, you eat local, good, whole foods. And you leave the other 30 percent for vices, airports, stadiums, movies, eating out or at someone's home. Please don't be THAT person at the dinner table asking whether your host is serving organic wild or farm-raised—it's simply rude.

Unfortunately, in today's society of faster, bigger, bolder, everyone's got their thing, and most of them aren't good things. The key is that when you are faced with the opportunity to dig into that somehow very satisfying buffalo chicken dip and have no idea where that chicken came from mixed with processed cream cheese and spicy red chemicals in the middle of a party where friends have gathered for a celebration, avoid being the pain in the ass asking a whole bunch of questions that quite frankly are annoying and annoy people. Either opt out or indulge, knowing it is part of your 30 percent food rule on bad eating, and counter it with fresh food next time.

Progress, science, growth and yield were motivations to feed the masses but left us overfed and actually starving for real food. We are

Every third bite of food you eat comes from honeybees. Local honey prevents allergies too! *Courtesy Cayla Zahoran.*

Baaaaabaaa black sheep free on the range foraging and growing to feed and clothe us all—the way life should be. *Courtesy Cayla Zahoran.*

beginning to realize how fundamental it is to go back to the farm to reboot our food chain and consume local and organically grown food without petrochemicals to prevent this illness and obesity.

The oldest honorable profession—farming—needs to be reestablished for our health, the health of our children and theirs and to economically develop our local communities sustainably.

Stone Barns Center is one of my favorite places to connect to seed and soil. It is one of the innovators impacting awareness about the local food movement in the Hudson Valley.

Stone Barns is a nonprofit eighty-acre farm created to increase public awareness about healthy, seasonal, sustainable food and to train farmers in resilient, restorative farming techniques. It also happens to have one of the best restaurants in the world on its property: Blue Hill at Stone Barns, serving haute, contemporary farm-to-table food, with an emphasis on the farm's produce, overseen by chef extraordinaire and author of *The Third Plate*, Dan Barber.

The center is named for the stone barns that crop so majestically out of the bucolic landscape; they were commissioned by John D. Rockefeller Jr. as dairy barns and were originally part of Pocantino, the Rockefeller estate, but they fell into disuse in the 1950s. In the 1970s, David Rockefeller's wife, Peggy, a cow-loving farm girl at heart, launched a successful cattle-breeding program. The barns have been in glorious use since. Stone Barns Center for Food and Agriculture was developed in honor of Peggy Rockefeller by David Rockefeller, their daughter Peggy Dulany and James Ford, an associate. It officially opened to the public in May 2004.

Stone Barns experiments with a range of animals and vegetables, using both traditional and newfangled methods to sustainably raise produce, poultry and animals. About two hundred varieties of produce are grown year-round in the six acres of vegetable fields, their gardens and in a twenty-two-thousand-square-foot greenhouse. Chemicals and pesticides are eschewed in favor of compost (see page 66 for my gardening cheat sheet). Chickens, turkeys, geese, sheep, pigs and bees are all raised humanely, and the farm keeps both its animals and plants healthy by using a rotational grazing method.

When Stone Barns launched in 2004, it seemed revolutionary—maybe even nostalgic—but as we catch up with its pioneer thinking and practices, we know it is the future of food. Stone Barns not only in anticipated prescribing a better food message for America with roots

Chef Jeremy Charles and Chef April Bloomfield team up to "Play with Fire" on an open pit for a FarmOn! feast in the Hudson Valley in August 2014. *Courtesy Cayla Zahoran.*

in agriculture but, ten years later, is also implementing practices from its long history in farming to keep the tradition of good food and local economies in play. This practice leaves more processed food on the shelf and gets people connected to seed and soil, not expiration dates.

If you can't read it, don't eat it. That's generally my rule when looking at the ingredient list of anything I eat—especially as the serving size often makes no sense. The beautiful thing about eating direct from the farm is that you eat pure goodness from people who work hard to feed you tasty food with a purpose. Farmers are honest folks and won't trick you or dupe you with false claims or poison you knowingly purely for profits.

Speaking of slick and false—it's a risky business to publicly question the big business of food. But if we don't, what will happen to our food shed and food systems? What will happen to the health of our youth? What will happen to livelihoods on our family farms and the integrity of Mother Nature herself? Who will feed us?

Agriculture builds community and commerce. It is vital for nutrition and economic development, and it just tastes better. If you are having a hard time seeing the right way to eating well, keep it simple. Keep farmers farming and adopt some of my "Meet Your Farmer" food rules to jumpstart you on a path to shopping and eating good-for-you food.

Food Rules to FarmOn!

Every day buy one local ingredient: Whether milk, cheese, vegetable, fruit, meat or poultry, it will taste better, you will tell someone and it will positively impact your health, the health of your community and the planet. Dairy is an easy start—look for Cabot or Hudson Valley Fresh options or Applegate natural and organic meat sourced from family farms.

Eat in season: Shop farmers' markets or look for in-season local fruit and organically grown vegetables at the grocery store. If they come from down the road and don't have petrochemicals, they may look less than perfect, but they will have more nutrition and taste better, especially if you haven't had them all year long! Not sure where to start? Peruse my list of favorite farmers' markets in the resource guide in the appendix.

Meet your farmer: If you don't want to grow your own food (a great learning experience for kids!), then visit a local farmer and meet the person who will provide you with food. They will explain the differences in your food choices and sources and the benefits of buying direct.

Black Angus cattle graze at Sir William Farm in Columbia County. *Courtesy Cayla Zahoran.*

Meet your meat: Having a connection to your food and where it came from is a good thing. It's humane to animals and good for you in every way possible. Free-range grazing animals are packed with more protein and nutrients. Farm-fresh meat is what Michael Pollan calls "salad bar meat" for its superior nutritional content, and the slightly pricier hand-raised meat is well worth the extra expense, not least of all because you need smaller portions of it to feel satisfied and full. How do you meet your meat? A visit to the farm is the best way to understand how vital the connection between animal, land and your table. Meet your farmer and ask about meat and poultry CSA programs. They make financial sense too.

Use common sense: Think about everything you put in your mouth. Instead of thinking about instant flavor sensation or cravings, ask, how do I benefit from eating this food? If it is immediate gratification, you will pay for it somehow—typically in digestion, skin, mood, sex and sleep. Treat your body like you respect it, and eat food as fuel for energy, pure joy in flavor and longevity with good health.

Invest in your food choices: Put your mind and money where your mouth is! If health were as revered as wealth, you would invest in food choices differently. If you can't eat local or organic, get informed about what you eat and opt out of processed food 70 percent of the time. Join a CSA and save. Ask your farmer to work with you on pricing and invest in the locavore lifestyle. Nothing is more important than your health—and it all starts with good food.

Understand sustainability: Learn about land stewardship and terroir. Ban junk food made in a plant—not from a plant—and embrace the capacity to endure and prolong a tradition of honest food. It's the long-term maintenance of well-being. Eating local food that exists in a system that connects us to the land, water, air, animals, people and places while fostering community and creating commerce and compliance brings viable livelihoods and profitability to farms, elements that are essential for small and mid-size family farms to survive and thrive. Succession is our future in farming. Get involved. Give the opportunity for our farmers (average age of fifty-eight years old!) to raise their replacements with rights to a future in the business of food.

Chapter 8

ACCOUNTABILITY

Opt Out of Processed Food

Imagine—we live in a world where food manufacturers spend more on the design and packaging than they do on the food intended for sale and consumption. They even advertise a famous brand of cookies on school report cards, so parents buy them to treat kids for good grades with junk food!

Food and beverage companies spend $2 billion every year enticing children with licensed cartoon characters in the form of toys and animation on food as advertising so kids convince their parents to buy crap to eat. Why don't they put those characters on all of the best food for you in the supermarket? Why don't we beat them at their own game and use compelling tactics to convince kids healthy is yummy and cool? The fast-food industry shells out a budget of more than $5 million daily to peddle its junk to kids, the Interagency Working Group on Foods Marketed to Children reports. Why don't they feed hungry children from local farms instead with that money?

Children watch on average more than ten food ads daily, 98 percent of which are for items that are high in fat, sugar or sodium. (As we know, Farmer Joe from down the street is just trying to get by, never mind bankrolling ads for his wholesome goodness.)

Parents get tricked just trying to "treat" their kids to something "fun" to eat, but it's a vicious cycle that leads to a lifetime of eating poorly. Almost 40 percent of calories in kids' diets comes from unhealthy fats and sugar, and only 21 percent of kids between the ages of six and nineteen eat five or more servings of fruit and vegetables a day.

Sickening, right?

Indeed. It seems to me that most of the big company's advertising dollars that you unwittingly support are spent purposefully undermining your family's health.

Interactive video games, a growth area for junk food advertising, are expected to see $1 billion worth of unhealthy food ads in 2014. About six million three- to eleven-year-olds are estimated to visit some sort of virtual game each month.

Groups like the Interagency Working Group on Foods Marketed to Children have proposed legislation and guidelines that would curtail some of the food industry's worst marketing tactics, but so far, all of the measures have failed. And according to the Center for Disease Control, if our junk food consumption trends continue, one in three American adults will have diabetes by 2050, and by 2030, healthcare costs that are related to bad diets could cost as much as $956 billion, or 17.6 percent of our healthcare costs.

How do we turn this around? Invest in real food from real farmers on clean land that you trust to feed you well. Plant a seed, touch the soil and get closer to real food.

Milk and its substitutes are one of the most confusing categories to me. Soy milk? Almond milk? Conventional dairy? Organic dairy? We know that the big business of food has the big money to create marketing claims that make it convenient for you to feel good about your food choices, but that doesn't necessarily make the food good, and all milk is not created equal.

Super milk comes from the super family farm practicing sustainable agriculture and bringing joy to your table from their hard work. Fresh local milk tastes better and is better for you. When you buy local food, you not only get more nutrition that makes you look and feel great, but you also better the environment, contribute to commerce and support your community, simply by connecting the dots from barn to belly.

Cheap food is cheating our health, and this is especially true with milk. Choose premium-quality local milk—you get what you pay for. Spending money on good food supports your health, supports the family farm and creates rural-to-urban marketplaces vital to economic development and viable livelihoods to change the future of farming.

If you think I'm obsessed with the future of milk in the Hudson Valley, you're correct. Dairy is the number one agricultural business in New York

State, and the state is number three in the country, behind California and Wisconsin, in overall dairy production. In the United States, the dairy industry is responsible for about $140 billion in economic output, $29 billion in household earnings and more than 900,000 jobs, according to Dairy Farming Today.

What's more important, though, is the "ripple effect" that buying one gallon of milk in your community provides. If a dairy farmer sells $1 worth of milk, it creates $3 worth of economic activity. For every $1 million in milk sales, seventeen jobs are generated. Dairies don't just create jobs for milkmen; they create jobs for veterinarians, insurance agents, farmers, truckers, salespeople and the micro-local economy of restaurants, gas stations, convenience stores, machine repair shops, etc. that get business from dairy farmers and their employees.

And you know that Greek yogurt everyone everywhere is eating these days? Much of it is produced in New York State. There are about forty yogurt plants in New York, and Governor Andrew Cuomo is giving grants for better equipment and providing other incentives to get dairy producers in the state to increase herd size and meet the growing demand. In 2014, Governor Cuomo made nearly $21 million available to state dairy farmers, and he is promising to help "make New York the Yogurt Capital of the nation."

It already is, according to the USDA. New York has been the top-ranked yogurt producer in the country since 2012, and we intend to keep it that way. In 2013, the dairy manufacturing industry alone employed about 9,470 people making a total of $513 million, up from 7,759 people earning $400 million in 2010, the USDA reports.

And what benefits one sector of farming benefits others as well. Agrana Fruit, one of the top producers of fruit preparations in the dairy industry, constructed a $50 million new manufacturing and distribution plant in 2014. It has already created more than sixty jobs and is set to employ about 120 people. It aims to process yogurt using New York–grown fruit. In addition to fruit farmers, it is helping the stainless steel and construction industries as well.

But as we've learned, with financial success comes greater risk. Dairy farmers in New York are finally enjoying some much-deserved stability, but it's important to stay true to our roots and not sacrifice overall sustainability for a quick buck. I'm not so worried about that when I look at the local dairy producers in my Hudson Valley neighborhoods.

Farmer Jeremy Peele treats his animals with respect and his land responsibly and feeds his community honest food beyond organic. *Courtesy Cayla Zahoran.*

The truth about quality lies in the comfort and cleanliness of the cows that the farmers carefully care for every day before your day even starts—and they do it because they love it, not to get rich. So why not show your love for them by drinking local milk every day (like Ronnybrook Farm Dairy in an iconic glass milk bottle) in your coffee or tea, in your cereal or soup or anytime you reach for that protein-rich deliciousness? Ask, where did it come from? What went into making my milk? What am I willing to pay to ensure quality? And how long did it take to get to my fridge? Is the milk you drink from cow to carton in thirty-six hours? Or do you trust that the word "organic" means fresh and nutrient dense—never mind that it is ultra-pasteurized? It's a great starting point to changing your diet—easy to do, and the impact is far-reaching.

Just a few years shy of celebrating the century mark of their old family farm, the Skodas of Triple Creek Farm make dairy farming look easy and feel like a dream. They treat a day at "the office" like any other executives in business, but they are in the dairy business in the open air and on their land seven days a week.

Value-added products in the Hudson Valley support local agriculture and economic development. *Courtesy Cayla Zahoran.*

"Agriculture is still New York State's number one revenue-generating business. We want people to remember that," farmer Richard Skoda reminded me.

Ann Pyburn Craig wrote about this issue in an article titled "Happy Cows Make Happy Milk: A Collaboration Between Hudson Valley Fresh and FarmOn!":

"New York State's rural economies are dependent on the survival of the dairy industry," said the state comptroller's office in a 2010 report. "These farms provide economic benefits, preserve open space, and are an integral part of rural life."

That report, entitled "New York's Dairy Farms In Crisis," paints a bleak picture of the viability of this crucial sector. In the latter half of the last decade, milk prices tanked dramatically, and traditional "milk cooperatives" were paying farmers $11.50 per hundredweight for milk that cost them at least $17.00 per hundredweight to produce. The comptroller's report mentions price supports, a pittance in USDA aid that "did not compensate for the tremendous losses suffered," and something called the Herd Buyout Program, which was discontinued in 2010 just as its proponents—dairy giants like Dairylea and Land O'Lakes, under the cozy moniker Cooperatives Working Together— faced increasing scrutiny over price-fixing and needless slaughter.

The damage has been deep. Another report, this one by the nonprofit Local Economies Project, concludes that the Hudson Valley has lost 70 percent of its dairy farms since 1987. Dr. Simon was motivated to start a farm by his experiences in both farming and medicine. "As an orthopedic surgeon, I saw people in their thirties with bones you'd expect in people in their seventies," he says, "and it comes down to lack of dairy. There is no calcium supplement you can take in old age that makes up for an early deficiency. In dairy farming, cleanliness is everything," says Simon. "There are farms with forty cows that are s^&*holes, and farms with four thousand cows that do a good job on that." Milk cleanliness is measured via testing for somatic cell count—somatic cells are the white blood cells that develop in response to bacteria—and HV Fresh holds its producers to a standard of under 200,000 per milliliter, less than a third of what the USDA considers allowable and less than half of what you'll find at most organic and traditional dairies.

The difference doesn't stop there. Dairy cows like HVF get a varied diet featuring lots of hay, which—partly because of the hay itself, and partly because of the thorough chewing it encourages—makes the milk higher in omega-3 fatty acids. No artificial bovine hormones are allowed. The milk is pasteurized simply and not "ultra-pasteurized" at the scalding temperature of 280 degrees Fahrenheit, a process bigger dairy operations use that sacrifices flavor and nutrition for longer shelf life and "safety" concerns (that are only an issue when there are higher somatic cell counts)...

In return for the exacting labor of cleaning and milking the five thousand cows that are their collective responsibility, Hudson Valley Fresh pays farmers like Richard Skoda a price that covers the cost of production. When milk cost $17.00 a hundredweight to produce, and the big co-ops were paying $11.50, Hudson Valley Fresh paid farmers $20.00.

"It should tell you something," observes Simon, "that even the New York State Department of Corrections, with free labor and tax-exempt land, decided they couldn't afford to run a dairy anymore."

Richard Skoda; his wife, Melissa; and their four children, Ryan, Joshua, Rachel and Alex, have dedicated their lives to making a difference by feeding us. They are sustaining the family tradition on their fourth-generation dairy farm in Taghkanic, New York, producing award-winning "super milk" for Agrimark products and Hudson Valley Fresh as a dairy of distinction growing grain for feed, caring for their cows fastidiously and maintaining equipment to make the farm run profitably.

Richard's father, Joe Skoda, was a dairy farmer, but he also dabbled in raising food for the family: hogs, meat chickens and laying hens. His grandfather John Skoda bought the forty-acre farm in 1922 in the same location where all of the family still lives today on the five hundred acres they have acquired since.

The family that farms together stays together, despite a tragic total loss from a fire in August 2006 when the family watched their barn—and their herd—burn at 3:00 a.m. The story brought a sadness to my heart for a loss so intense that it inspired me to share their story of fearlessness to rebuild their business at a time when most would have given up farming.

The Skodas are indeed food entrepreneurs, and their success isn't counted in numbers but measured by time. The "Getirdone" spirit at Triple Creek Farm offers perseverance and dedication I can only describe as passion.

Richard and his family designed and rebuilt the two hundred–herd, free stall–style barn at Triple Creek Farm today with a succession

plan and skill set for their next generation to heed the calling to the land and operate as a dairy farm in the Hudson Valley for another century to come.

Brothers Ryan and Joshua Skoda are capable, fearless young farmers, too—smart, funny, friendly, knowledgeable, compassionate and driven in the community. They are guaranteed to succeed. They love what they do, and it shows. And you can support them too just by buying local milk.

Never mind a day off or making a fortune—these are businessmen who rise and shine ready to work milking cows, birthing calves, growing feed on one thousand acres of farmland to feed their Holstein herd and even raising Guernsey breeds for their kids, Brayden, Lydia, Bella and Mazie.

Their mantra is quality, and together they told me that superior "super milk" can be achieved by feeding animals well and treating them well. The goal at this dairy is always having clean, happy cows that don't suffer from stress in any form. Not only do happy cows produce tastier milk (demonstrated by the accolades Hudson Valley Fresh gets every year), but they "let down" more product: each Skoda cow produces nearly eighty-two pounds of milk per day.

"We believe that quality brings you a good investment on your return, it's good for overall herd health and it's good for the bottom line. It's good for making money. Quality is number one around here. We keep our standards high, we keep our barns clean, our cows clean, and we also believe in cow comfort," Richard Skoda explained.

When I stopped by for the afternoon milking at 4:00 p.m.—twelve hours after the farmers' days started—the herd of two hundred "girls" lined up for a visit to the parlor for milking that makes you wonder if farmers are really "cow whisperers." The milking parlor is state-of-the-art and run by just family members and one full-time employee.

Timing was perfect. When I arrived, a calf was born (there are about 120 young stock on the farm), and the pride Ryan greeted me with to show off his new "girl" made me feel honored to be invited to Triple Creek Farm.

I asked Ryan if I could name her. He explained you had to give her a name matching the first letter of her mom Chloe and follow these rules: no names repeated (he keeps a book), she could not be named after anyone we knew and her name would include the prefix Skoda Grove followed by the sire name. So I named her Chelsea because I could tell

Opposite: A perfect autumn afternoon in the Hudson River Valley. *Courtesy Tessa Edick.*

from her shiny black coat and one white spot that she was a city girl with heart. Skoda Grove Windhammer Chelsea was ready for a debutante walking just as soon as she was birthed on her toes!

And, who knew, but apparently there's a whole lot of city-like living for these girls—"manicures" twice yearly with nail clippings, "fluffed" beds with fresh sand for better footing and mobility and fresh local food on which to gorge themselves, filling their bellies and then lounging in comfort, regurgitating their food to chew their cud some forty to sixty times (Ryan said he counts), reveling in happiness in between milkings at the parlor twice daily.

Mooooooooooooo!

Since all cows are female, there is no surprise they are finicky, and they don't like change. If you even change their feed by tweaking it, they can smell it and turn up their noses in a fuss. Cows want the exact same food the same way at the same time every day. If you even change it at all, they won't eat it. They demand a routine and don't really like newcomers. They like what they know, which is why a diet of hay and grazing is so satisfying. The cost of grazing is so much higher that corporations won't sacrifice profits to make grass-fed milk a widespread offer on supermarket shelves. They consider it unsustainable because of the time and grassland needed to feed America's voracious appetite for dairy and meat. (Joel Salatin, one of my favorite farmer-scribe-philosophers, speaking about grass-fed beef, stated that "eco-ag can feed the country and the world" if we simply collectively agree to change our ways and our commitment to corn and stop letting the associations, fertilizer and agribusiness industries telling us it can't be done.)

I'm asking: why not change? Why aren't the corporate farms able to deliver consistent care for the animals that produce milk we are all convinced is good for you to drink?

Honest and true, all of this care and comfort leads to one thing: quality milk. As dairy cooperative farmer Richard Skoda told me, "I don't think about the public saying thank you—we don't expect that. I am just thankful I see my children and my grandchildren almost every day. We want farming to stay farming in this area—if we don't farm, farmer services will go away, and it will be hard to keep going. I'm always thinking about efficiencies—grease and maintenance keep everything going."

On the farm, you need everything to work: the equipment, the family, the animals, the soil, the air and the water. It's vital to success and quality at its best.

When it comes to cheese, being a black sheep isn't always a bad thing.

In a land that celebrates four cheeses, cheese melts and commodity cheese, it's good to stand up for wholesome goodness and opt out of any and all processed cheese choices permanently.

It's a story we all want to buy into. It's about choosing locally made food, meeting your farmer and making a big difference with small choices like cheese—every meal, every day. Start asking: was the cheese made with dairy? Is it good for my body and my planet?

Tom and Nancy Clark of Old Chatham Sheepherding Company in Old Chatham, New York, designed and built their Shaker-style farm on sixty acres of land in the heartland of agriculture in the Hudson Valley. This working farm dates back to 1935 and is as bucolic and gracious as any family farm I know. You can buy any of their delicious yogurt flavors or three types of cheese direct from the farm—Camembert (Nancy's Wheel or Square, a mix of cow/sheep milk), Kinderhook (pure sheep milk) or Blue (Pasteurized Ewe's Blue or Unpasteurized Shaker Blue),

Old Chatham Sheepherding Company is a family farm that has integrated into retail markets, the Cornell School of Agriculture, a farmstead creamery and advanced technology in cheese making. *Courtesy Cayla Zahoran.*

all made with their famed black sheep milk from the herd they raise and twenty-five employees who make it possible.

Married for fifty years as of 2013, Mr. and Mrs. Clark and their three children have farming in their hearts and Cornell University on their résumés, having all graduated from this land-grant college in Upstate New York. They still look to Cornell for consulting and advice to improve their offers today as a trusted source for information on all aspects of life sciences and agriculture. Tom told me Cornell is a farmer's friend indeed and explained his motivation: "Being involved and building a good team that shares and executes ideas is a business that is well run, feels good and is well respected in the marketplace." According to its website, the Cornell Small Farms Program mission is to "foster the sustainability of diverse, thriving small farms that contribute to food security, healthy rural communities, and the environment. We do this by encouraging small farms–focused research and extension programs and fostering collaboration in support of small farms."

Looking back, the success of the cheese seems to have been written in the stars. Their recipe combined savvy graphics with delicious cheese products and produced an instantly beloved line of cheeses recognized, coveted and loved by consumers. (Old Chatham's infamous green packaging with a silhouetted black sheep was designed by Tom's brother-in-law Ivan Chermayeff, an iconic graphic designer behind instantly recognizable logos such as those for PBS, Barneys and NBC.)

Sheep are sheep. Black or white, it is simply a matter of genetics. The black breed at Old Chatham Sheepherding Company are 100 percent east Friesian and a more hardy flock. Even the great-great-granddaughters grazing the grounds today are a mix of black Friesian ewe and a white ram, but the black genes dominate and survive. Visually, the red barns and black free-range sheep on grassy green pastures make a visit to the farm dreamy.

Speaking of dreams, Tom Clark grew up on a farm in Arlington outside of Poughkeepsie, and by ten years of age, he had raised three sheep, which he took to the Dutchess County Fair and won first prize showing the breed. He told me, "I developed a love for animals, and my grandpa let me do whatever I wanted. I drove a tractor at twelve through the hay fields and had a love for farming ever since."

This led to an education at Cornell College of Agriculture, with studies in animal husbandry and economics and, after graduation, a career in private equity, in which he is still invested today. But Tom's passion for farming was already imprinted, and in 1993, when the farm became available with a residence, Tom decided to take a leap of faith and purchase the land. He went back to his roots and had five Cornell students take on a project to vet the idea of the area supporting a country inn and restaurant at the farm.

When the answer was yes, they found a local chef who had graduated first in his class at the Culinary Institute of America in Hyde Park, New York, and worked with Alice Waters to make the restaurant a home for Hudson Valley produce and eaters. (Alice Waters is a chef, author and the proprietor of Chez Panisse. She is an American pioneer of a culinary philosophy that maintains that cooking should be based on the finest and freshest seasonal ingredients that are produced sustainably and locally. She is a passionate advocate for a food economy that is "good, clean, and fair." Over the course of nearly forty years, Chez Panisse has helped create a community of scores of local farmers and ranchers whose dedication to sustainable agriculture assures the restaurant a steady supply of fresh and pure ingredients.) "We didn't know what we were doing but quickly gained a great reputation for food and wine, and just two years later, in 1995, there was a three-month waiting list for a Saturday-night table at our farm-to-table restaurant and B&B," said Tom Clark.

A Hudson Valley cheese course is a delightful way to entertain guests and start conversation about eating local. *Courtesy Cayla Zahoran.*

Even New York City critics came, ate and raved! This prompted the launch of a bakery and first-class treatment for all guests and food enthusiasts, with cheese courses for dessert and black sheep cookies appearing for turn-down service before bed at the B&B.

In 1999, the restaurant graciously closed so the Clarks could return their focus to what they did best: make award-winning cheese. And win they did. More than fifty awards line the tasting room with Specialty Food Trade Sofi Awards, the "Oscars" for food, and state fair first prizes in cheese making from various states and master cheese makers.

Consumers agreed, and the timing was right, which helped their products sell nationally to retailers like Whole Foods Market and Wegmans. This small family farm and creamery became a successful brand in the yogurt and cheese categories.

In New York City, Murray's Cheese collaborates with the farm, "taking young, fat wheels of sheep's milk cheese from Old Chatham Sheepherding in the Hudson Valley and giving them a Corsican accent, with a coating of rosemary, lemon thyme, marjoram, elderberries and hop flowers. The cheese, called Hudson Flower, then ages for about a month in Murray's caves," as Florence Fabricant reported in the *New York Times*.

It's that easy. Make food choices that are more than labeled healthy and fresh. Ask questions. Understand your role in fixing our broken food system. It tastes better, and it's easy to do. Imagine the change that will transpire from that one choice every day if we all simply opt out of cheesy alternatives.

Shop local and meet your farmer, like Tom Clark. You, too, will become obsessed with local farms and the return on investment they bring.

All you need is food, so why wouldn't you invest in what you eat to avoid being sick or fat? Be conscious and spend your food dollars wisely.

It saves you so much money in the long run paying it forward for good health with prevention. And in our agrarian society, it is easy. But you have to make that choice. Shift your lifestyle by eating with awareness and invest in a diet to work in harmony for health and prevention of illness with clean food. Connect to what you eat and where your food comes from instead of buying into a new set of labels marked "certified" and buzzing with ideas that we don't yet know whether they are worthy

Opposite: Free-range, pasture-raised, open-air chickens that live happy and healthy are the ONLY poultry you want on your plate. *Courtesy Cayla Zahoran.*

Meet your meat. British white heritage cattle are beautiful inside and out. *Courtesy Cayla Zahoran.*

of our ingestion or investment. Use common sense, un-complicate things and get back to the farm.

Skip the supermarket this harvest season, and on a crisp, sunny fall day, make the farm your weekend destination for good food. Open the proverbial barn door so that you can take a peek at the amazing smorgasbord of options. Touch a seed. Plant something you grow. Eat it and share it with others. Consume locally grown food. It serves us all better. It is responsible eating from sources that care. It's a sustainable system that can feed the next generation—and the masses, too. It fuels your body and mind today and stimulates economic development in communities for the long term. This is a victory in food!

On the farm, you will adore memorable walks, blue skies and fresh air, and your kids will have the time of their lives frolicking in the fields. You will leave knowing why farmers do what they do, and you will fall in love, too. Pick and choose what works for you, and leave the rest for someone else to try. It's a rewarding lifestyle. Change the way you eat. It starts with your food choices and sources. It's that simple. Start today and leave thinking the same: thank goodness someone told me that! FarmOn!

ONLINE RESOURCES

You've read the book. Now here's your cheat sheet! These are some of my favorite online sources for finding out more on implementing best practices for food shopping and consumption.

FarmAid.org

http://www.farmaid.org/site/c.qlI5IhNVJsE/b.2723725/k.8DCF/
    Food_Labeling.htm

FarmAid.org offers an excellent guide and resource center to good food sources, explaining different food labels to look for and which are certified, sources for accessing local food seasonally and a vast network of partners to educate us all on good food.

JustFood.org

http://justfood.org

JustFood.org empowers and supports community-led efforts to access locally grown food, especially in underserved neighborhoods. Just Food provides training and education and helps people launch farmers' markets and CSAs.

LocalHarvest.org

http://www.localharvest.org/farmers-markets/

Search for farmers' markets, family farms and other sources of sustainably grown food in your area where you can buy produce, grass-fed meats and many other goodies. If you can't find what you're looking for close to home, check out their catalogue to order good food online.

Know Your Farmer Know Your Food
http://www.usda.gov/wps/portal/usda/usdahome?navid=
    KNOWYOURFARMER
Learn how the USDA supports good food from family farmers and local and regional food economies. Explore the Know Your Farmer, Know Your Food Compass to find out what's happening near you.

JustLabelIt.org
The organization behind the legal petition calling for the mandatory labeling of genetically engineered foods (written by attorneys at the Center for Food Safety). Learn how to sign the petition and join the millions of signers and hundreds of partner organizations representing healthcare, consumers, farmers, parents and environmentalists.

Foodroutes.org
A national organization dedicated to building local food systems. Check out their Facebook and join the community and movement! https://www.facebook.com/FoodRoutes?ref=hl

Eatwellguide.org
A free online directory of family farms, restaurants, markets and other outlets of fresh, locally grown food throughout the United States and Canada.

Homegrown.org
A community for celebrating the culture of agriculture and sharing skills like growing, cooking and canning.

FARMERS' MARKETS

WINTER FARMERS' MARKETS

http://www.nyfarmersmarket.com/farmers-market-profiles/
markets/2013-2014-winter-markets.html
The Farmers' Market Federation of New York is a membership
organization of farmers' markets and their managers, sponsors, consumers
and supporters. It is a priceless database of farmers' markets info, including
when/where/how, plus it provides opportunities to volunteer.

BERKSHIRE COUNTY

Great Barrington Farmers' Market
40 Castle Street
Great Barrington, Massachusetts
(413) 528-8950
gbfarmersmarket.org

COLUMBIA COUNTY

Chatham Farmers' Market
15 Church Street
Chatham, New York
(518) 392-3353
chathamrealfoodcoop.net

Copake Hillsdale Farmers' Market
Roe Jan Park
Hillsdale, New York
(518) 329-0384
Saturday, 9:00 a.m.–1:00 p.m.

Hudson Farmers' Market
Sixth and Columbia Street
Hudson, New York
hudsonfarmersmarketny.com
Saturday, 9:00 a.m.–1:00 p.m.

Philmont Farmers' Market
116 Main Street
Philmont, New York
pbinc.org/revitalization
Sunday, 10:00 a.m.–1:00 p.m.

Upstreet Market Farmers' Market
Warren Street
Hudson, New York
https://www.facebook.com/
   upstreetmarket
Wednesday after work

DUTCHESS COUNTY

Amenia Farmers' Market
Amenia Town Hall parking lot,
   4988 Route 22
Amenia, New York
(845) 373-4411
ameniafarmersmarket.com
Friday, 3:00 p.m.–7:00 p.m.

Beacon Farmers' Market
8 Red Flynn Road, across from
   Beacon Train Station
Beacon, New York
(845) 234-9325
hebeaconfarmersmarket.com
Sunday, 11:00 a.m.–3:00 p.m.

Fishkill Farmers' Market
1004 Main Street
Fishkill, New York
(845) 897-4430
Thursday, 9:00 a.m.–3:00 p.m.

Hyde Park Farmers' Market
4383 Albany Post Road
Hyde Park, New York
(845) 229-9336
hydeparkfarmersmarket.org
Saturday, 9:00 a.m.–2:00 p.m.

LaGrange Farmers' Market
M&T Bank Plaza, 4 Jefferson
   Plaza
LaGrange, New York
(914) 204-0924
Friday, 3:00 p.m.–7:00 p.m.

Millbrook Farmers' Market
Tribute Garden, 3219 Franklin
   Avenue
Millbrook, New York
(845) 677-3697
millbrooknyfarmersmarket.com
Saturday, 9:00 a.m.–2:00 p.m.

Millerton Farmers' Market
Railroad Plaza, Main Street
Millerton, New York
(518) 789-4259
millertonfarmersmarket.org
Saturday, 9:00 a.m.–1:00 p.m.

Pawling Farmers' Market
Charles Colman Boulevard
Pawling, New York
http://pawlingfarmersmarket.org
Saturday 9:00 a.m.–12:00 p.m.,
   June–September

Rhinebeck Farmers' Market
Municipal Parking Lot, 61 East
  Market Street
Rhinebeck, New York
(845) 876-7756
www.rhinebeckfarmersmarket.com
Sunday, 10:00 a.m.–2:00 p.m.

ORANGE COUNTY

Chester Farmers' Market
Winkler Place
Chester, New York
(845) 476-6241
Sunday, 9:00 a.m.–3:00 p.m.

Cornwall Farmers' Market
Town Hall, 183 Main Street
Cornwall, New York
(845) 534-9100
cornwallcoop.com
Wednesday, 10:00 a.m.–4:00 p.m.;
  Saturday, 10:00 a.m.–2:00 p.m.

Florida Farmers' Market
Route 17A and Route 94 Junction
Florida, New York
(845) 641-4482
Tuesday, 11:30 a.m.–5:30 p.m.

Goshen Farmers' Market
Village Square, Main and South
  Church Street
Goshen, New York
(845) 294-7741
Friday, 10:00 a.m.–5:00 p.m.

Middletown Farmers' Market
Erie Way from Grow to Cottage
  Streets
Middletown, New York
(845) 343-8075
Saturday, 8:00 a.m.–1:00 p.m.

Monroe Farmers' Market
Museum Village, 1010 Route 17M
Monroe, New York
(845) 344-1234
Wednesday, 9:00 a.m.–3:00 p.m.

Montgomery Farmers' Market
Clinton Street
Montgomery, New York
(845) 616-0126
Saturday, 9:00 a.m.–2:00 p.m.

Newburgh Farmers' Market
Downing Park, Route 9W and
  South Street
Newburgh, New York
(845) 565-5559
Friday, 10:00 a.m.–4:00 p.m.

Newburgh: Healthy Orange
  Farmers' Market
131 Broadway, between Lander
  and Johnston Streets
Newburgh, New York
(845) 568-5247
Tuesday, 10:00 a.m.–3:00 p.m.

Newburgh Mall Farmers' Market
Parking lot, 1401 Route 300
Newburgh, New York
(845) 564-1400
Saturday, 10:00 a.m.–2:00 p.m.

Pine Bush Farmers' Market
Corner of Main and New Streets
Pine Bush, New York
(845) 978-0273
pinebushfarmersmarket.com
Saturday, 9:00 a.m.–2:00 p.m.

Walden Farmers' Market
1 Municipal Square
Walden, New York
(845) 476-6241
Friday, 11:30 a.m.–4:30 p.m.

Warwick Farmers' Market
Corner of South and Bank Streets
Warwick, New York
(845) 222-5947
warwickvalleyfarmersmarket.org
Sunday, 9:00 a.m.–2:00 p.m.

West Point Farmers' Market
Municipal Parking Lot, Main
  Street
West Point, New York
(917) 509-1200

PUTNAM COUNTY

Cold Spring Farmers' Market
Boscobel House and Gardens, 1601 Route 9D
Cold Spring, New York
csfarmmarket.org
Saturday, 8:30 a.m.–1:30 p.m.

ULSTER COUNTY

Gardiner Farmers' Market
Gardiner Library, 133 Farmer's
  Turnpike
Gardiner, New York
(845) 255-1255
Friday, 4:00 p.m.–8:00 p.m.

Kingston Midtown Farmers'
  Market
Broadway, between Henry and
  Cedar Streets
Kingston, New York
(347) 276-2606
rket.org
Tuesday, 3:00 p.m.–7:00 p.m.

Kingston Uptown Farmers'
  Market
303 Wall Street
Kingston, New York
(845) 853-8512
www.kingstonfarmersmarket.org
Saturday, 9:00 a.m.–2:00 p.m.

Milton Farmers' Market
Cluett-Schantz Park, 1801–1805
  Route 9W
Milton, New York
(845) 616-7824
hhvfarmersmarket.com
Saturday, 9:00 a.m.–2:00 p.m.

New Paltz Farmers' Market
Il Gallo Giallo parking lot, 36
  Main Street
New Paltz, New York
(845) 255-5995
newpaltzfarmersmarket.com
Sunday, 10:00 a.m.–3:00 p.m.

Rosendale Farmers' Market
Rosendale Community Center,
  1055 Route 32
Rosendale, New York
(845) 658-3467
rosendalefarmersmarket.com
Sunday, 9:00 a.m.–2:00 p.m.

Saugerties Farmers' Market
115 Main Street
Saugerties, New York
(845) 246-6466
saugertiesfarmersmarket.com
Saturday, 10:00 a.m.–2:00 p.m.

Woodstock Farmers' Market
6 Maple Lane
Woodstock, New York
(845) 679-5345
woodstockfarmfestival.com
Wednesday, 3:30 p.m.–dusk

WESTCHESTER COUNTY

Peekskill Farmers' Market
Bank Street
Peekskill, New York
(914) 737-2780
Saturday, 8:00 a.m.–2:00 p.m.

White Plains Farmers' Market
Court Street between Martine
  Avenue and Main Street
White Plains, New York
Wednesday May–November, 8:00
  a.m.–4:00 p.m.
(914) 422-1336

HUDSON VALLEY FARM-TO-TABLE RESTAURANTS

Allium Restaurant
44 Railroad Street
Great Barrington, Massachusetts
(413) 528-2118

Another Fork in the Road
1215 Route 199
Milan, New York
(845) 758-6676

A Tavola Trattoria
46 Main Street
New Paltz, New York
(845) 255-1426
www.atavolany.com

Babette's Kitchen
3293 Franklin Avenue
Millbrook, New York
(845) 677-8602

Backyard Cooking Company
Germantown, New York
http://backyardcookingcompany.com
info@backyardcookingcompany.com

Blue Hill at Stone Barns
630 Bedford Road
Tarrytown, New York
(914) 366-9600
http://www.bluehillfarm.com/food/
    blue-hill-stone-barns

Bywater Bistro
419 Main Street
Rosendale, New York
(845) 658-3210
www.bywaterbistro.com

Café le Perche
230 Warren Street
Hudson, New York
(518) 822-1850

Café Mio
2356 Route 44
Gardiner, New York
(845) 255-4949
www.miogardiner.com

Chatham Brewing
59 Main Street
Chatham, New York
(518) 697-0202
http://www.chathambrewing.com

Culinary Institute of America:
    American Bounty Restaurant
Route 9 (1946 Campus Drive)
Hyde Park, New York
(845) 471-6608
www.ciarestaurants.com

The Dancing Cat Saloon and
    Catskill Distilling Company
2037 Route 17B
Bethel, New York
(845) 583-3141
www.dancingcatsaloon.com
www.catskilldistilling.com

Depuy Canal House
1315 Main Street
High Falls, New York
(845) 687-7700

The Farmers Wife
3 County Route 8
Ancramdale, New York
(518) 329-5431
http://www.thefarmerswife.biz

Fish & Game Restaurant
13 South Third Street
Hudson, New York 12534
(518) 822-1500

Global Palate
1746 Route 9W
West Park, New York
(845) 385-6590
www.globalpalaterestaurant.com

The Greens
44 Golf Course Road
Copake Lake, New York
(518) 325-0019

Harney & Sons Tea Shop
13 Main Street
Millerton, New York
(518) 789-2121

Henry's at the Farm at Buttermilk
    Falls Inn
220 North Road, Milton
(845) 795-1500
www.henrysatbuttermilk.com

The Hop
458 Main Street
Beacon, New York
(845) 440-8676
www.thehopbeacon.com

Il Gallo Giallo
36 Main Street
New Paltz, New York
(845) 255-3636

Irving Farm Coffee House
44 Main Street
Millerton, New York
(518) 789-2020

Jar'd Wine Pub
10 Main Street
New Paltz, New York
(845) 255-8466

John Andrews Restaurant
224 Hillsdale Road
Great Barrington, Massachusetts
(413) 528-3469

No. 9
53 Main Street
Millerton, New York
(518) 592-1299
http://number9millerton.com

Old Mill
53 Main Street
Egremont, Massachusetts
(413) 528-1421
http://oldmillberkshires.com

Peekamoose Restaurant & Tap
   Room
8373 State Route 28
Big Indian, New York
(845) 254-6500
www.peekamooserestaurant.com

The Phoenicia Diner
5681 Route 28
Phoenicia, New York
(845) 688-3175
http://www.phoeniciadiner.com

Red Devon
108 Hunns Lake Road
Bangall, New York
(845) 868-3175
www.reddevonrestaurant.com

Serevan
6 Autumn Lane
Amenia, New York
(845) 373-9800

Stissing House
7801 South Main Street
Pine Plains, New York
(518) 398-8800
http://www.stissinghouse.com

Sweet Grass Grill
24 Main Street
Tarrytown, New York
(914) 631-0000
www.sweetgrassgrill.com

Swoon Kitchen Bar
340 Warren Street
Hudson, New York
(518) 822-8938
http://www.swoonkitchenbar.com

The Tavern at Diamond Mills
25 South Partition Street
Saugerties, New York
(845) 247-0700
www.diamondmillshotel.com

Tavern at the Highlands Country
   Club
955 Route 9D
Garrison, New York
(845) 424-3254
www.highlandcountryclub.net

Terrapin
6426 Montgomery Street
Rhinebeck, New York
(845) 876-3330
www.terrapinrestaurant.com

Tuthill House at the Mill
20 Gristmill Lane
Gardiner, New York
(845) 255-4151
www.tuthillhouse.com

• APPENDIX 3

Valley at the Garrison
2015 Route 9
Garrison, New York
(845) 424-3604
www.thegarrison.com

HISTORICAL HOMES GUIDE

Estates of the Hudson Valley

H udson Valley estates were built by unofficial members of the American aristocracy. Their estates (and sometimes farms) live on today as beautiful historical relics of the period in which they were built, showcasing the best period architecture, landscape design and interior decoration money could buy. Most estates were built during the late nineteenth or early twentieth centuries and feature Federal period details and/or Revival styles. Many of them are open to public visitors and have been maintained and restored.

They are more than restorations, of course: they provide a backdrop and context for the people and events that shaped our amazing country—presidents, robber barons, farmers, war heroes and authors of the Declaration of Independence all lived here in the Hudson Valley. Come see the past and glimpse the future at some of their old haunts.

BOSCOBEL (GARRISON)

Van Cortlandt Manor was purchased by John D. Rockefeller Jr. in the 1940s. The stone manor, built in 1732, is flanked by a rebuilt tavern and a restored tenant house. Expect live demonstrations of period activities including cooking, spinning, weaving and brickmaking. Tours of the

manor by costumed guides include many original period furnishings and a spacious kitchen with a traditional open hearth and beehive oven.
Riverside Avenue, Croton-On-Hudson, New York (914) 271-8981, or contact Historic Hudson Valley at (914) 631-8200. Admission fee.

VANDERBILT MANSION (HYDE PARK)

Built by the third generation of Vanderbilts in the Neoclassical style in 1899, the estate is a great example of the wealth and excess of the Gilded Age. From the columned porch at the rear of the mansion, you will catch magnificent glimpses of the Hudson River as it roars by. Several species of old trees grace the grounds, and formal gardens on the property have been recently restored to their former splendor.
519 Albany Post Road, Hyde Park, New York (845) 229-9115, http://www.nps.gov/vama. Admission fee.

WILDERSTEIN (RHINEBECK)

Built originally in the Italian villa style, Wilderstein was remodeled to a Queen Anne in the 1880s. Its circular tower soars five stories above a landscape created by noted American Romantic landscape artist Calvert Vaux. Stained-glass pieces by J.B. Tiffany grace the library. The last member of the Suckley family to call Wilderstein home was a cousin and lifelong confidante of Franklin D. Roosevelt. Her papers and memorabilia, along with those of her family, are available for perusal.
Morton Road, PO Box 383, Rhinebeck, New York (845) 876-4818, http://www.wilderstein.org. Admission fee.

KYKIUT (SLEEPY HOLLOW)

A Hudson Valley landmark, this hillside six-story castle was home to four generations of Rockefellers, beginning with John D. Find terraced gardens; twentieth-century sculptural treasures from Picasso, Henry Moore, Alexander Calder and David Smith; underground galleries with Picasso tapestries; and a cavernous coach barn with classic automobiles and horse-drawn carriages. Guided tours.
381 North Broadway, Sleepy Hollow, New York.

Sunnyside (Tarrytown)

For horror and history fans, Washington Irving's Sunnyside is a must-see where visitors can learn about the inspiration behind "The Legend of Sleepy Hollow" and characters like Brom Bones and Ichabod Crane. Guides dress in period attire and walk visitors through the quirky design of Sunnyside, which echoes colonial New York homes and cottages in Scotland and Spain. The grounds, creeping wisteria and water features add to its charm and make the perfect backdrop for picnics. Admission fee.

Philipsburg Manor (Sleepy Hollow)

The manor, built in the 1700s, was a thriving farming, milling and trading center for the Philipses, who also rented out the land to tenant farmers and relied on slave labor to operate. Visitors are thrown back into eighteenth-century life and can explore the too-often glossed-over realities of enslavement in the colonial period in New York. The manor house—with its dairy, kitchens, bedchambers, parlors, slave gardens, textile production centers and warehouses—is all intact. Admission fee.

Montgomery Place (Annandale on Hudson)

With 380 acres of formal gardens, ancient trees, rolling hills and arboretums, an estate, perennial gardens, orchards, waterfalls and innumerable other features, it would be easy to spend a full day here. The mansion includes Classical Revival exteriors designed by Alexander Jackson Davis. A farm stand offers produce from the orchards in season. Admission fee.

Clermont (Germantown)

Seven members of the Livingston family lived here, including Robert, who was one of the five people who wrote the Declaration of Independence. He also swore in George Washington as America's first president. His first mansion was lit on fire by invading British troops in 1775. It was rebuilt and then remodeled in the 1920s as a Colonial Revival. There

is an impressive collection of portraits and sculptures and many special events, including croquet tournaments and antique shows.
1 Clermont Avenue, Germantown, New York (518) 537-4240. Admission fee.

GLENVIEW (YONKERS)

This Victorian mansion built in 1877 is part of the Hudson River Museum of Westchester complex, which also boasts the Hudson River Museum and the Andrus Planetarium. It is considered a prime example of Eastlake interior styling, with extensive stenciling and woodwork that are inspired by nature.
511 Warburton Avenue, Yonkers, New York (914) 963-4550. Admission fee.

LINDENWALK (KINDERHOOK)

Martin Van Buren, the eighth president of the United States, was born in Kinderhook and purchased this estate in 1839, remodeling it in a Federal and Italianate Revival style. The estate boasts a large collection of historic wallpaper and Hudson Valley archaeological ephemera.
1013 Old Post Road, Kinderhook, New York. Admission fee.

LOCUST GROVE (POUGHKEEPSIE)

The inventor of the telegraph and the Morse code, Samuel F.B. Morse, purchased the estate in 1847 and converted it into a Tuscan villa with the help of architect A.J. Davis. Period interiors and a Morse Exhibition Room, with a copy of the original telegraph model, are on display. Roughly 150 acres of wildlife, walking trails and amazing river views await.
370 South Road, Poughkeepsie, New York (845) 454-4500. Admission fee.

LYNDHURST (TARRYTOWN)

Like a Gothic castle, this 1838 estate hovers over the Hudson, with turrets, battlements and a soaring tower. Mayor of New York City and general William Paulding commissioned architect A.J. Davis to construct the

Greek Revival fortress. Subsequent owners made additions to the home (including a four-story tower) and the grounds (a greenhouse and aviary). 635 South Broadway, Tarrytown, New York (914) 631-4481.

STAATSBURGH MILLS MANSION (STAATSBURGH)

Gilded Age society mavens Ogden Mills and Ruth Livingston built the sixty-five-room estate around an inherited mansion in 1895. It showcases Beaux Arts Neoclassical styling and fantastical seventeenth- and eighteenth-century English and French furnishings.
Old Post Road, Staatsburg, New York (845) 889-8851. Admission fee.

OLANA (HUDSON)

Hudson River School of Art painter Frederic Church's Persian palace is a work of art in and of itself. It reflects Church's travels in the Middle East and Europe, underpinned by his love of the Hudson Valley. Paintings from Church and the rest of the members of the Hudson River School of Art are joined by artifacts from his travels; exterior landscaping in the Romantic style completes the holistic celebration of near and far.
5720 Route 9G, Hudson, New York (518) 828-0135. Admission fee.

SPRINGWOOD (HYDE PARK)

Franklin D. Roosevelt, America's thirty-second president, was born here, lived here for much of his life and was buried here, along with his wife, Eleanor, in a beautiful rose garden. The mansion was built in the Georgian Colonial style in the early 1800s. Formal busts of FDR and sculptures abound in the landscape, which is perfect for strolling and picnicking. There is also an FDR Library and Museum on site, with many fascinating historical documents, special programs and tours, as well as the FDR Home Garden and educational Victory Garden.
519 Albany Post Road, Hyde Park, New York (845) 229-9115. Admission fee.

VAL-KILL (HYDE PARK)

A Dutch Colonial cottage built for Eleanor Roosevelt, it is a streamside spot on the larger Springwood estate grounds. It was created in 1926 and became a sanctuary for the first lady and the president during hectic times. Kruschchev, Winston Churchill and Haile Selassie all spent time here.
519 Albany Post Road, Hyde Park, New York (845) 229-9115. Admission fee.

FARM-TO-SCHOOL PROGRAMS

This article by Anne Pyburn Craig, "Happy Cows Make Happy Milk: A Collaboration Between Hudson Valley Fresh and FarmOn!" appears on Country Valley Wisdom (http://www.countrywisdomnews.com/2014/09/happy-cows-make-happy-milk_2.html).

This year, kids in three eastern Hudson Valley school districts will be drinking fresh, top-quality, local milk with their lunches, thanks to a confluence of creative energy and funding from several sources—and the urging of the kids themselves.

"Many of our students were drinking Hudson Valley Fresh milk at home," says Sandra J. Gardner, public information officer at Taconic Hills Central Schools in Columbia County's Craryville. "This initiative really got started when they started asking for better tasting milk to be served in our school lunch program."

Integrating local supply chains with major institutions that practice economies of scale is notoriously difficult. Policies requiring contracts be awarded to the low bidder are hard to challenge in tough economic times. But the kids, parents, and health-oriented staff of Taconic Hills—a school that has its own garden and greenhouse—had a strong ally in Tessa Edick, a wellness committee member whose résumé put her in the perfect position to help.

Edick is a dynamic redhead with boundless energy and a gift for making connections that is surpassed only by her culinary chops. She is the founder of Culinary Partnership and two nonprofits, Friends of the Farmer and FarmOn! Foundation, which she describes as the natural outgrowth of her career in gourmet production and marketing. Prior to creating the partnership, she led her company Sauces 'n Love Inc. to 13 Sofi awards, "the Oscars of specialty foods," as she puts it, and millions in revenues.

Now Edick, a relocated New Englander, has devoted her evangelical passion and gift for glamour to the Hudson Valley locavore scene. Taconic Hills kids wanted Hudson Valley Fresh, and Tessa Edick leveraged her various partnerships to make it happen.

The milk the kids have gotten so fond of is produced by a nine-farm dairy cooperative founded in 2005 by former state assemblyman Patrick Manning and Dr. Sam Simon, a onetime Middletown dairy farm kid who'd spent 22 years doing orthopedic surgery before going back to the farm—just at the moment when many were leaving it for good.

"New York State's rural economies are dependent on the survival of the dairy industry," said the state comptroller's office in a 2010 report. "These farms provide economic benefits, preserve open space, and are an integral part of rural life."

That report, entitled "New York's Dairy Farms In Crisis," paints a bleak picture of the viability of this crucial sector. In the latter half of the last decade, milk prices tanked dramatically, and traditional "milk cooperatives" were paying farmers $11.50 per hundredweight for milk that cost them at least $17 per hundredweight to produce. The comptroller's report mentions price supports, a pittance in USDA aid that "did not compensate for the tremendous losses suffered," and something called the Herd Buyout Program, which was discontinued in 2010 just as its proponents—dairy giants like Dairylea and Land O'Lakes, under the cozy moniker Cooperatives Working Together—faced increasing scrutiny over price-fixing and needless slaughter.

The damage has been deep. Another report, this one by the nonprofit the Local Economies Project, concludes that the Hudson Valley has lost 70% of its dairy farms since 1987. Dr. Simon was motivated to start a farm by his experiences in both farming and medicine. "As an orthopedic surgeon, I saw people in their 30s with bones you'd expect in people in their 70s," he says, "and it comes down to lack of dairy. There is no calcium supplement you can take in old age that makes up for an early deficiency."

Hudson Valley Fresh began with two farms, one of them Simon's own, and since has grown to nine, located throughout Columbia, Dutchess, and Ulster counties. Standards are strict. "In dairy farming, cleanliness is everything," says Simon. "There are farms with 40 cows that are s^&*holes, and farms with 4,000 cows that do a good job on that." Milk cleanliness is measured via testing for somatic cell count—somatic cells are the white blood cells that develop in response to bacteria—and HV Fresh holds its producers to a standard of under 200,000 per milliliter, less than a third of what the USDA considers allowable and less than half of what you'll find at most organic and traditional dairies.

The difference doesn't stop there. HV Fresh cows get a varied diet featuring lots of hay, which—partly because of the hay itself, and partly because of the thorough chewing it encourages—makes the milk higher in omega-3 fatty acids. No artificial bovine hormones are allowed. The milk is pasteurized simply and not "ultra-pasteurized" at the scalding temperature of 280 degrees Fahreheit, a process bigger dairy operations use, which sacrifices flavor and nutrition for longer shelf life and "safety" concerns (that are only an issue when there are higher somatic cell counts).

Beyond that, HV Fresh promises consumers "cow to carton in 36 hours." That part of the process takes place at the family-owned Boice Brothers Dairy in Kingston, where they recently celebrated their first century in the business by using over a ton of ice cream to craft the world's longest ice cream sundae.

HV Fresh dairy products meet kosher standards and are beloved by gourmet Manhattan coffee shops, where the quality of the milk in the latte is what keeps 'em coming back. It's no shock that kids who were used to it at home began to lobby for it at school; "Like Breyers' chocolate ice cream in a glass," is how one satisfied customer describes HV Fresh chocolate milk.

In return for the exacting labor of cleaning and milking the 5,000 cows that are their collective responsibility, HV Fresh pays farmers a price that covers the cost of production When milk cost $17 a hundredweght to produce, and the big co-ops were paying $11.50, HV Fresh paid farmers $20.

"It should tell you something," observes Simon, "that even the New York State Department of Corrections, with free labor and tax-exempt land, decided they couldn't afford to run a dairy anymore."

Good nutrition in school cafeterias has gained enormous momentum with a federal mandate, and Edick decided FarmOn! resources would be well expended in helping districts bridge the gap between the cost of conventional milk and the HV Fresh product the kids preferred. Taconic Hills wrote a health and wellness addendum to district bylaws that allowed them to deviate from standard purchasing procedure, and FarmOn! did the rest.

"The vendor we used previously would charge us $0.27 a carton," explains Gardner, "and this coming year, Hudson Valley Fresh will charge us $0.30 per carton of fat free chocolate milk. Farm On! Foundation has agreed to supplement the 3 cent differential. This is the second year Farm On! Foundation has supported the school's initiative to support local milk consumption. Our a la carte consumption has increased 18.7% from the 2012/2013 to 2013/2014 school year." Edick and Dr. Simon say the district's overall milk consumption is up 300 percent.

FarmOn! has sponsoring partners, among them Whole Foods, the state government (via TasteNY), and the Local Economies Project of the New World Foundation, along with a few smaller private companies; the organization collaborates with Cornell on educational programs. But Edick's gift for lending farming the glamour that she feels it deserves makes her a powerhouse fundraiser in and of herself. An event entitled "Play With Fire" on August 17 brought 10 celebrity chefs and 6 acclaimed mixologists to the fields of Fish and Game Farm in Hudson, where they prepared locally sourced delicacies and libations for guests who paid $250 apiece and up for the pleasure. Earlier this summer, a Hootenanny at the Copake Country Club brought in over $60,000 for the nonprofit's cause: making sure that the next generation understands food.

Taconic Hills is currently one of three school districts receiving the milk subsidy. Edick plans to expand that number to 15, and get HV Fresh milk onto SUNY campuses as well. "Tessa's a powerhouse," says Dr. Simon. "Farmers are lucky she came to live here; so are the schoolkids." That slurping sound you hear? Thousands of milk cartons being emptied to the dregs.

HUDSON VALLEY'S DRINKING TRAIL

Has there ever been a better time to drink in New York State? Governor Andrew Cuomo has infused the beverage production industry with a $6 million marketing and promotional campaign, and he has overseen the loosening of Draconian regulations left over from Prohibition that significantly reduce business costs, allowing a delicious boozy renaissance to flourish here, only fitting for the home of America's oldest winery, established in 1839: Brotherhood Winery. (Since 2011, there has been an 83 percent growth in farm-based beverage licenses.)

New York is home to more than six hundred wineries, breweries, distilleries and cideries. It ranks third in the nation for wine and grape production and second in apple production, has the second most distilleries and boasts three of the twenty top producing breweries in the country. I am not going to list all of them here, but I will point you toward a handful of my favorites in each category, most of which are open to visitors (check their websites first).

COLUMBIA COUNTY

Chatham Brewing, LLC
59 Main Street
Chatham, New York
www.chathambrewing.com

Harvest Spirits
3074 U.S. Route 9
Valatie, New York
www.harvestspirits.com

Hillrock Estate Distillery
408 Pooles Hill Road
Ancram, New York
www.hillrockdistillery.com

Hudson-Chatham Winery
1900 State Route 66
Ghent, New York
www.hudsonchathamwinery.com

Tousey Farms and Winery
1783 Route 9
Germantown, New York
www.touseywinery.com

DUTCHESS COUNTY

Alison Wines
231 Pitcher Lane
Red Hook, New York
www.alisonwines.com

Cascade Mountain Winery &
  Restaurant
835 Cascade Mountain Road
Amenia, New York
www.cascademt.com

Hyde Park Brewing Company
4076 Albany Post Road
Hyde Park, New York
http://www.hydeparkbrewing.
  moonfruit.com/#

Millbrook Winery, Inc.
26 Wing Road
Millbrook, New York
www.millbrookwine.com

Mill House Brewing Company
289 Mill Street
Historic Poughkeepsie, New York
(845) 485-2739

ALBANY COUNTY

Nine Pin Ciderworks, LLC
929 Broadway Avenue
Albany, New York 12207
http://www.ninepincider.com/

ULSTER COUNTY

Adair Vineyards
52 Alhusen Road
New Paltz, New York
www.adairwine.com

Baldwin Vineyards
176 Hardenburgh Road
Pine Bush, New York
www.baldwinvineyards.com

Benmarl Winery & Slate Hill
   Vineyards
156 Highland Avenue
Marlboro, New York
www.benmarl.com

Keegan Ales
20 Saint James Street
Kingston, New York
http://www.keeganales.com

Nostrano Vineyards
14 Gala Lane
Milton, New York
www.nostranovineyards.com

Robibero Winery
714 Albany Post Road
New Paltz, New York
www.rnewyorkwine.com

Stoutridge Vineyard
10 Ann Kaley Lane
Marlboro, New York
www.stoutridge.com

Tuthilltown Spirits
14 Gristmill Lane
Gardiner, New York
www.tuthilltown.com

Whitecliff Vineyard
331 McKinstry Road
Gardiner, New York
www.whitecliffwine.com

SENECA COUNTY

Swedish Hill Vineyard, Inc.
4565 State Road 414
Romulus, New York
www.swedishhill.com

SCHENECTADY COUNTY

Helderberg Meadworks
PO Box 93
Duanesburg, New York

SULLIVAN COUNTY

Catskill Distilling Company
2037 Route 17B
Bethel, New York
www.catskilldistilling.com

GREENE COUNTY

Crossroads Brewing Company
21 Second Street
Athens, New York
crossroadsbrewingco.com

ORANGE COUNTY

Applewood Winery
82 Four Corners Road
Warwick, New York
www.applewoodwinery.com

Brotherhood Winery
100 Brotherhood Plaza Drive
Washingtonville, New York
www.brotherhood-winery.com

Clearview Vineyard
35 Clearview Lane
Warwick, New York
www.clearviewvineyard.com

Demarest Hill Winery & Distillery
81 Pine Island Turnpike
Warwick, New York
Demaresthillwinery.com

Newburgh Brewing Company
88 South Colden Street
Newburgh, New York
http://www.newburghbrewing.com/

Palaia Vineyards
10 Sweet Clover Road
Highland Mills, New York
www.palaiavineyards.com

Rushing Duck Brewing Company
1 Battiato Lane
Chester, New York
http://www.rushingduck.com/

Warwick Valley Winery & Black
    Dirt Distillery
114 Little York Road
Warwick, New York
www.wvwinery.com
www.blackdirtdistillery.com

Westtown   Brew   Works   and
    Hopyard
236 Schefflers Road
Westtown, New York
http://westtownbrewworks.com/

WESTCHESTER COUNTY

Captain Lawrence
444 Saw Mill River Road
Elmsford, New York
http://www.captainlawrencebrewing.com/

Peekskill Brewery
47–53 South Water
Peekskill, New York
http://thepeekskillbrewery.com/

BIBLIOGRAPHY

Books

Pollan, Michael. *Food Rules: An Eater's Manual.* New York: Penguin Books, 2009.

The Prince of Wales HRH. *The Prince's Speech: On the Future of Food.* With a foreword by Wendell Berry and afterword by Will Allen and Eric Schlosser. New York: Rodale, 2012.

Studies, Online Resources, Articles

Special note regarding my column, "Meet Your Farmer," in the *Register-Star* (http://www.registerstar.com/columnists/meet_your_farmer/): Many of the conversations, background information and interviews that appear in this book were documented either in my column or in the research for my column. Specific columns are not cited here but are available in the archives of the *Register-Star* online.

"Agrana Fruit Formall Opens Lysander Plant." May 15, 2014. http://www.cnybj.com/News/Articles/TabId/102/ArticleId/39218/language/en-US/agrana-fruit-formally-opens-lysander-plant.aspx#.VCrFqVYVzwI.

"The Business Case for the Green Economy: Sustainable Return on Investment." http://www.unep.org/greeneconomy/Portals/88/

documents/partnerships/Executive%20Summary_Business_case. pdf, United Nations Environment Programme report.

Center for Disease Control and Prevention. "Adult Obesity Facts." http://www.cdc.gov/obesity/data/adult.html.

———. "Chronic Disease Prevention and Health Promotion." http://www.cdc.gov/chronicdisease/resources/publications/aag/ddt.htm.

Chism, J.W., and R.A. Levins. "Farm Spending and Local Selling: How Do They Match Up?" *Minnesota Agricultural Economist*, 1994. Cited in Sustainable Table's Food Economics: http://www.sustainabletable. org/491/food-economics.

Dairy Farming Today. "State Statistics." http://www.dairyfarmingtoday. org/Learn-More/FactsandFigures/Pages/StateStatistics.aspx.

———. "The U.S. Dairy Industry: A Vital Contributor to Economic Development." http://www.dairyfarmingtoday.org/SiteCollectionDocuments/economicfactsheet.pdf.

"Global Consumers Are Willing to Put Their Money Where Their Heart Is When It Comes to Goods and Services from Companies Committed to Social Responsibility." June 17, 2014. http://www.nielsen.com/us/en/press-room/2014/global-consumers-are-willing-to-put-their-money-where-their-heart-is.html.

Glynwood. "The State of Agriculture in the Hudson Valley." http://www.glynwood.org/files/2011/02/State_of_Ag_2010.pdf.

"Governor Cuomo Announces New York State Is Top Yogurt Producer in Nation." May 27, 2014. https://www.governor.ny.gov/press/05272014-nys-top-yogurt-producer.

"Growth Patterns in the U.S. Organic Industry." October 24, 2013. http://www.ers.usda.gov/amber-waves/2013-october/growth-patterns-in-the-us-organic-industry.aspx#.VCq4P1YVzwI.

Howard, Philip H. (associate professor, Michigan State University). "Organic Processing Industry Structure." https://www.msu.edu/~howardp/organicindustry.html.

"Improving Childhood Nutrition, New Gillibrand Report: Nearly 60 Percent of New Yorkers Overweight." http://www.gillibrand.senate.gov/agenda/improving-childhood-nutrition.

"Interagency Working Group on Food Marketed to Children." http://www.iaapa.org/docs/gr-archive/110428foodmarketproposedguide.pdf?sfvrsn=2.

"New Data Reflects the Continued Demand for Farmers Markets." August 4, 2014. http://www.usda.gov/wps/portal/usda/usdahome?contentid=2014/08/0167.xml.

New York State Office of the State Comptroller. "New York's Dairy Industry Is in Crisis." https://www.osc.state.ny.us/localgov/pubs/research/snapshot/0310snapshot.pdf.

The Old Farmer's Almanac. "2014 Best Planting Dates Calendar for Albany, NY." http://www.almanac.com/gardening/planting-dates/NY/Albany.

Prevention Institute. "The Facts on Junk Food Marketing and Kids." http://www.preventioninstitute.org/focus-areas/supporting-healthy-food-a-activity/supporting-healthy-food-and-activity-environments-advocacy/get-involved-were-not-buying-it/735-were-not-buying-it-the-facts-on-junk-food-marketing-and-kids.html.

"Report: Organic Industry Achieved 25 Years of Fast Growth Through Fear and Deception." April 22, 2014. http://www.foodsafetynews.com/2014/04/report-fast-growing-organics-industry-is-intentionally-deceptive/.

"United States Organic Food Market Forecast & Opportunities, 2018." March 25, 2014. http://www.prnewswire.co.uk/news-releases/united-states-organic-food-market-forecast--opportunities-.

"U.S. Department of Agriculture's Net Farm Income Forecast to Fall in 2014." http://www.ers.usda.gov/topics/farm-economy/farm-sector-income-finances/highlights-from-the-2014-farm-income-forecast.aspx#.VCrAClYVzwI.

"U.S. Environmental Protection Agency's Crop Production." http://www.epa.gov/oecaagct/ag101/printcrop.html.

SPEECH

Eric Schlosser's speech in 2011 for the conference On the Future of Food held at Georgetown University. Link to excerpted remarks: http://www.washingtonpost.com/lifestyle/food/eric-schlosser-on-the-future-of-food/2011/05/06/AF5x1siG_story.html.

INDEX

ABOUT THE AUTHOR

A creative idea machine and constant source of high energy, Tessa Edick brings her passion and big ideas for foodies to the Hudson Valley farming community in New York.

Edick is the founder and executive director of FarmOn! Foundation, a 501(c)(3) nonprofit organization and public charity committed to raising the awareness of the Hudson Valley farming community and educating the public on the importance of local farms. FarmOn! offers programs for adults and youth teaching them how to incorporate farm-fresh ingredients into their daily lives—and help support their community in the process.

An experienced food entrepreneur with twenty-plus years of successful food launches behind her, Edick expresses her creativity with innovative brand strategies, business development, networking and memorable social marketing concepts.

Edick also is the founder of the Culinary Partnership, where she contributes her expertise to help celebrity chefs and entrepreneurs launch their recipes to retail shelves. The popular food products launched by Culinary Partnership are in line with her commitment to farm-fresh ingredients and help promote differentiation and responsible food choices.

Edick launched her career in the food industry as the founder of Sauces 'n Love, Inc., a Boston-based fifteen-year-old food manufacturer for whom she developed two brands of Italian specialty food products and turned them into multiple multimillion-dollar revenue streams. Edick revolutionized ready-made sauce in a jar and single-handedly secured more than two hundred print and television media impressions as she built the company into an international force. She has earned sixteen NASFT Sofi Awards (the "food Oscars"), and her products were featured twice in *Oprah Magazine*'s "O List."

Edick's newspaper column, "Meet Your Farmer," appears bimonthly in the Hudson Valley's Columbia-Greene Media newspapers. She currently resides in Copake, New York, with her boyfriend and English bulldog, Ms. Ruby Juice.